New and Improved . . .

The Science Reference Library

New and Improved...

Inventors and inventions that have changed the modern world

R. Baker

Published for the British Library
by British Museum Publications Limited

On 1st July 1973 the Library collections formerly belonging
to the Trustees of the British Museum were transferred to
the ownership of the British Library Board.

© 1976, The British Library Board
ISBN 0 7141 0380 2
Published by British Museum Publications Ltd.
6 Bedford Square, London WC1B 3RA

Designed by Harry Green
Set in 11/12½ point Photon Times
and printed in Great Britain by
The Camelot Press Ltd, Southampton

Contents

Note on Frontispiece

The frontispiece overleaf shows an example of a Letters Patent Deed for a patent held by the National Research Development Corporation (NRDC) which was set up by the Government in 1949 to ensure that full and proper use be made of British inventions. The NRDC has since lent its financial support to a number of widely different products and processes including such well-known inventions as the Cockerell hovercraft and the Bacon fuel cell.

Cephalosporin C is a non-toxic antibiotic, effective against a wide range of bacteria, including some which are resistant to the traditional penicillins. It was developed after intensive research carried out by a British research team on cultures originally isolated from a sewage outfall in Cagliari, Sardinia, by Professor Giusseppe Brotzu.

Patent No. 810196

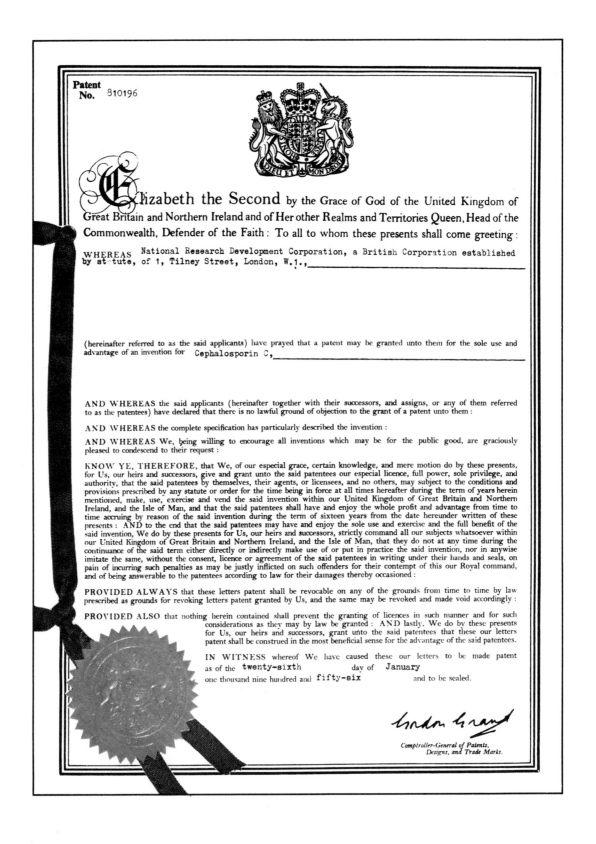

Elizabeth the Second by the Grace of God of the United Kingdom of Great Britain and Northern Ireland and of Her other Realms and Territories Queen, Head of the Commonwealth, Defender of the Faith: To all to whom these presents shall come greeting:

WHEREAS National Research Development Corporation, a British Corporation established by statute, of 1, Tilney Street, London, W.1., _____

(hereinafter referred to as the said applicants) have prayed that a patent may be granted unto them for the sole use and advantage of an invention for Cephalosporin C, _____

AND WHEREAS the said applicants (hereinafter together with their successors, and assigns, or any of them referred to as the patentees) have declared that there is no lawful ground of objection to the grant of a patent unto them:

AND WHEREAS the complete specification has particularly described the invention:

AND WHEREAS We, being willing to encourage all inventions which may be for the public good, are graciously pleased to condescend to their request:

KNOW YE, THEREFORE, that We, of our especial grace, certain knowledge, and mere motion do by these presents, for Us, our heirs and successors, give and grant unto the said patentees our especial licence, full power, sole privilege, and authority, that the said patentees by themselves, their agents, or licensees, and no others, may subject to the conditions and provisions prescribed by any statute or order for the time being in force at all times hereafter during the term of years herein mentioned, make, use, exercise and vend the said invention within our United Kingdom of Great Britain and Northern Ireland, and the Isle of Man, and that the said patentees shall have and enjoy the whole profit and advantage from time to time accruing by reason of the said invention during the term of sixteen years from the date hereunder written of these presents: AND to the end that the said patentees may have and enjoy the sole use and exercise and the full benefit of the said invention, We do by these presents for Us, our heirs and successors, strictly command all our subjects whatsoever within our United Kingdom of Great Britain and Northern Ireland, and the Isle of Man, that they do not at any time during the continuance of the said term either directly or indirectly make use of or put in practice the said invention, nor in anywise imitate the same, without the consent, licence or agreement of the said patentees in writing under their hands and seals, on pain of incurring such penalties as may be justly inflicted on such offenders for their contempt of this our Royal command, and of being answerable to the patentees according to law for their damages thereby occasioned:

PROVIDED ALWAYS that these letters patent shall be revocable on any of the grounds from time to time by law prescribed as grounds for revoking letters patent granted by Us, and the same may be revoked and made void accordingly:

PROVIDED ALSO that nothing herein contained shall prevent the granting of licences in such manner and for such considerations as they may by law be granted: AND lastly, We do by these presents for Us, our heirs and successors, grant unto the said patentees that these our letters patent shall be construed in the most beneficial sense for the advantage of the said patentees.

IN WITNESS whereof We have caused these our letters to be made patent as of the twenty-sixth day of January one thousand nine hundred and fifty-six and to be sealed.

Comptroller-General of Patents,
Designs, and Trade Marks.

Introduction

Every week the United Kingdom Patent Office receives some thousand or more applications for patents for invention. Each of these must, if it is to lead to the grant of a valid patent, lay claim to some product or process of manufacture, or an improvement to an existing product or process, which was not previously known in the United Kingdom. Each application must be supported by a document, known as a specification, which must describe the invention sufficiently for any competent person to copy the product or apply the process, and must also set out precisely the inventive steps that the inventor claims as his own. The application will in due time be examined by one of the Patent Office examiners, who will search through his file of existing patent specifications to make sure that the inventor's claims do not conflict with those of earlier inventors. He will also check certain other statutory requirements, for example, that the descriptions of the invention and of the extent of the claims are clearly understandable and that the subject matter of the invention is in fact patentable. It would not be patentable, for example, if it was frivolous, contrary to law or morality, if it was a food or medicine made from a mixture of known ingredients or if it were a mere form of words. This last provision has been held to rule out computer programs and the like.

If all the statutory requirements are met then the application will be accepted by the Patent Office and the specification will be published. Normally this publication takes place about two years after the filing of the first application. After publication there is a further period of three months before the patent is granted and formally sealed. During this period any interested party may, if he considers he has good cause, oppose the grant. Such opposition often leads to amendments being made to the wording of the specification to circumvent the objections made. Even after the grant, the patent is still open to challenge, but the legal procedures differ.

By 1971 the overall number of published United Kingdom patent specifications had reached well over one and a half million. The present rate of acceptance of new applications is about 800 per week or over 40,000 per year, so that the specifications form a substantial and rapidly growing library of technological information. Even so this United Kingdom contribution is only a small part of the total output of patent information. All the world's major industrial countries have some form of patenting system. In many the output of printed specifications is greater than that of the United Kingdom: the weekly output of printed accepted specifications from the United States is 1,500 to 1,600, West Germany print around 1,200 patent applications per week and the Japanese over 2,000. One of the commercial indexing services[1] which extended its activities during 1974 to cover all published patent specifications from twenty-four of the major patenting countries anticipated that this would involve the handling of 12,000 specifications per week. This does not represent so many new inventions. In order to obtain patent protection in any particular country it is almost invariably the case that a separate patent has to be filed in that country. An inventor who hopes to obtain the maximum profit from his invention will apply for patent protection in all those countries where there may be some worthwhile market potential. Thus, overall, there will be more patent specifications than there are new inventions. The number of allegedly new inventions being protected by patent applications is still impressive. Recent estimates give a world figure of 400,000 per year, or 1,600 per working day.

The majority of these new inventions will not have any marked impression on the development of science or technology as a whole. Most will relate to minor improvements in a product or process and will give the patentee some commercial advantage for a time. Some may have no commercial value in that, despite the high hopes of their inventors, there is no way of developing the ideas put forward to the stage of profitable application. Others will fail to produce a significant impact or to become commercially viable because there is no entrepreneur with sufficient business expertise, energy and resources who is willing to back the invention through its initial teething troubles. Together with all these there will be an occasional specification describing an invention destined to have an immense and lasting effect, perhaps leading to the creation of a complete new industry or giving fresh vigour to an existing one, thereby changing social habits and patterns in the years to come.

The list which forms the main part of this book marks an attempt to select, from the vast literature of patent specifications, some of those relating to the more significant inventions. Many of the inventions referred to would be quickly recognised by the public at large and the names of the inventors be well known. For many others, although the invention may concern products in common use, the significant inventive steps may not have been so widely publicised and the inventions would not therefore be familiar outside the industry concerned.

The list had its origins in a small collection of references built up over the years by the staff of the British patents enquiry desk of the Science Reference Library. Additions to this collection had been entered from time to time whenever a reader's enquiry had led to the discovery of some patent that seemed worth recording. This collection has been augmented for publication. A search has been made through a large number of textbooks, encyclopaedias and journal articles for references to significant inventions, and the patents relating to them have been traced wherever possible. In addition, members of the Patent Office examining staff have suggested patents which, in their view, are significant. The author is grateful for the assistance that has been given in this way and for the general encouragement by the Patent Office in the compilation of this list.

The purpose of all patent systems has always been to encourage inventive persons to continue to invent, to put their inventions to work and to pass on their knowledge and skills to the ultimate benefit of the state. This philosophy was well set out in the wording of the first known grant of a monopoly patent for an industrial invention. This grant, for a three-year period, was made in 1421 by the Italian state of Florence to Filippo Brunelleschi, the famous engineer and architect responsible for the building of some of the architectural treasures of Florence, including the Palazzo Pitti and the Cathedral dome. The invention concerned a barge with special hoisting gear which was to be used to transport marble. The grant contained the following wording:

'Because Brunelleschi did not want to give the invention to public use for fear of being robbed of the reward of his labours, the privilege is granted with the express intention not only that the invention may be made useful as well for himself as for the generality but particularly also that he himself may be urged to further exertion, and stimulated to achieve greater inventions; the Government agrees to protect the inventor against unauthorised working and to grant the author an immediate monopoly for the period stated by prohibiting the use of every form of transport ship not in use at the date of the privilege unless it be built by Brunelleschi himself or with his consent.'

This concept of the advantage to the state to be accrued in return for limited period monopolies was reiterated in the first known patent law enacted in Venice in 1474. A modern exposition of the same principles can be found in the report published in 1970 by the Banks Committee set up to examine the United Kingdom Patent System and Patent Law:[2]

'We have found a general acceptance that the act of invention and the development of new ideas is inherent in the human mind and would continue without any legal protection for the results. As, however, a patent system increases the possibility of reward for the successful exploitation of invention, there can be little doubt that it does play a part in encouraging individuals to invent and organisations to create conditions in which inventions can be made. But the basic aim of a patent system, and indeed its effect, is to encourage the successful industrial application of inventions. The man with the resources can normally be expected to put those resources to industrial use without specialist assistance in established fields, where he can be reasonably assured that his factory will work technically and where the demand for his product is known to exist. If, however, resources are to be put at risk to develop a new process or product, which has yet to be tested, then he will hesitate lest the expense of the development may yet prove to be irrecoverable while his competitors can wait and, without equivalent expense, pick up and use the successful results. It is the knowledge that a patent monopoly will enable him to hold off competition for a period that encourages him to take the risk and use those resources to develop new industrial inventions.'

When an inventor is granted *letters patent* he is given a monopoly to make, use or sell his invention for a fixed period of time; in the United Kingdom this is sixteen years. The inventor is required to pay fees at various times to maintain his monopoly. Nowadays these fees do little more than pay for the expenses of running the system—paying for the publication of the specifications, the employment of staff, the maintenance of buildings and so on. The State's return for the monopoly grant comes indirectly from the commercial benefits derived in a more rapid distribution of technological knowledge and a greater willingness by businessmen to incur the risks of breaking new ground.

The published specifications present a vast storehouse of technological knowledge which can be consulted openly by anyone with an interest. Information gleaned from it can be used freely by anyone once the period of the monopoly comes to an end. Many patents expire before the full sixteen years because the patentee stops paying renewal fees. From the specifications, interested parties, possibly better equipped than the patentee to exploit the invention, can learn about the invention and have the opportunity of approaching the patentee with a view to arranging a licensing agreement. In any event the information can be used for research purposes by anyone who is not applying such use to direct commercial gain.

Much of the information contained in the specifications may never be published elsewhere since it may be of too detailed a nature or be considered too abstruse or speculative to attract the editors of journals or the publishers of books. Many ideas are described in patent specifications long before they become sufficiently well known or acceptable to appear in other forms of literature. A glance at the entry in this book for the jet propulsion engine will serve to illustrate this point. Whittle's ideas on jet propulsion were described in patent specifications well before the Second World War, while the basic concept of using a jet to propel aircraft had appeared many years earlier.

Scientists, technologists, historians and others who might benefit from the information contained in patent specifications often appear to have a marked reluctance to consult them.

There are several reasons for this. First, there are not enough libraries with substantial holdings of patents. A number of public reference libraries in the major industrial centres throughout the country do have substantial holdings of British patent specifications and many of these also take specifications from some overseas countries (pp. 27–8); and many major industrial companies or research associations have libraries with significant holdings of specifications covering their particular interests. But this still leaves a substantial body of readers without any convenient reference collection. Moreover there are still many reference librarians unaware of the availability, or of the potential value, of patent specifications.

A second reason for the uninitiated to hesitate in the use of patent literature is that the majority of the specifications are written using a special legalistic jargon often referred to as 'patentese'. Most inventors employ patent agents to write their specifications. These agents need to be specially knowledgeable both in the technical subject matter of the invention and also in the legal requirements of the patent law. Large firms employ agents directly within their own patent departments while smaller firms and private inventors may make use of firms of patent agents.[3] Agents justify the use of patentese by pointing out that the patent may have to be defended in court and the extent of the inventor's protection may hinge on the particular form of words used. This is particularly true about that part of the specification containing 'the claims' which, according to Blanco-White[4] in his well-known book on British patent law, should be 'at least as carefully drafted as sections in an Act of Parliament or clauses in a will'. There is therefore a tendency to use words and phrases which have been tested in court and will, without ambiguity, express precisely what is meant without in any way being limiting in their meaning so as to give a competitor a chance to circumvent the patent. So the phrase 'a plurality of' will be used to denote 'more than one', 'a member' to describe what in actual fact is a wooden strut but which might be replaced by a metal bar or some suitably shaped piece of rigid plastic, 'measuring apparatus' instead of instrument, etc.

This discipline, imposed by the requirements of the law, has advantages as well as disadvantages for those who come to patent specifications looking for information. One important requirement is that the specification should sufficiently describe the invention so that anyone who is reasonably knowledgeable in the subject matter of the patent ('practised in the art' is the patentese) should be able to copy the invention without further assistance. This means that patents usually contain more detail on constructional matters or chemical constituents than might be given in other forms of literature. There is a recognised order in which the various component parts of a specification are arranged. Anyone who studies patent specifications is soon able to recognise this common structural pattern and can quickly locate the component parts. In order of occurrence, there is a formal opening paragraph containing the name and address of the patentee and a request for the grant of a patent; a statement of the subject matter of the invention; an account of the 'prior art' (this is patentese for what is known already and is not always included in United Kingdom specifications); a statement of the purpose of the invention (this too may not be included although the patent can be invalidated at any time if it can be shown that the invention it relates to serves no useful purpose), the 'consistory clause' which sets out the ambit of the invention claimed; an explanation about any drawings that have been included (the drawings themselves are normally on separate sheets at the end of the specification); a detailed description of the invention, usually including 'examples' or 'embodiments' of the invention; and, finally, a detailed numbered list of the claims.

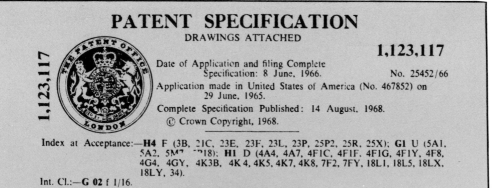

PATENT SPECIFICATION
DRAWINGS ATTACHED

1,123,117

1,123,117

Date of Application and filing Complete
Specification: 8 June, 1966.

No. 25452/66

Application made in United States of America (No. 467852) on
29 June, 1965.

Complete Specification Published: 14 August, 1968.

© Crown Copyright, 1968.

Index at Acceptance:—**H4** F (3B, 21C, 23E, 23F, 23L, 23P, 25P2, 25R, 25X); **G1** U (5A1, 5A2, 5M7, 2P18); **H1** D (4A4, 4A7, 4F1C, 4F1F, 4F1G, 4F1Y, 4F8, 4G4, 4GY, 4K3B, 4K 4, 4K5, 4K7, 4K8, 7F2, 7FY, 18L1, 18L5, 18LX, 18LY, 34).

Int. Cl.:—**G 02** f 1/16.

COMPLETE SPECIFICATION

Electric field device

We, WESTINGHOUSE ELECTRIC CORPORA-TION, a Corporation organised and existing under the laws of the Commonwealth of Pennsylvania, United States of America, of
5 Three Gateway Center, P.O. Box 2278, Pittsburgh 30, Pennsylvania, United States of America, do hereby declare the invention, for which we pray that a patent may be granted to us, and the method by which it
10 is to be performed, to be particularly described in and by the following statement:—
This invention relates to electric field responsive devices for the detection and visual indication of electric fields.
15 In many applications, it is desirable to provide an electric field sensitive device for giving a visual indication of the fact that a field is present and also visual indication in response to a change or different values
20 of the field thereacross. Certain organic materials have been found that modify their optical properties in response to an electric field. These materials are liquid crystalline materials of the cholesteric phase.
25 Liquid crystalline materials have properties that are intermediate to those of a true liquid and a true crystal. It has mechanical properties of a liquid and optical properties of a crystal. Liquid crystal-
30 line materials have also been referred to as materials in the mesomorphic state.
Liquid crystals are conventionally divided into three classes or phases. One class is the smectic structure, which is characterized
35 by its molecules being arranged in layers with the long axis of the molecules approximately normal to the plane of the layers. A second class is the nematic structure, which is characterized by threadlike molecules that
40 tend to be in nearly parallel orientation with their long axis. The nematic phase does not have the molecules separated into layers. The third class of liquid crystals is known as the cholesteric phase. The
45 molecules in the cholesteric phase substantially act as though they are arranged in

layers. These layers are very thin with the long axis of the molecules parallel to the plane of the layers. It is also found that the direction of the long axis of the mole- 50 cules in each layer is displaced slightly from the corresponding direction in adjacent layers. The term cholesteric was chosen because the molecular structure is characteristic of a large number of compounds 55 that contain cholesterol. Cholesterol by itself does not have a liquid crystal phase. The present invention is concerned with materials exhibiting a cholesteric liquid crystalline phase. 60
The unique molecular architecture of the cholesteric liquid crystals give rise to a number of optical properties which differ from the smectic or nematic phase. The characteristic properties of the cholesteric 65 structure may be summarized as follows:
(1) It is optically negative, while smectic and nematic structures are optically positive. An optical negative layer provides that light entering perpendicular to the mole- 70 cular layers has a maximum velocity.
(2) The structure is optically active. If linearly polarized light is transmitted perpendicularly to the molecular layers, the direction of the electric vector of light would be 75 rotated progressively to the left along the helical path. Thus the plane of polarization, which is determined by the electric vector in the direction of propagation, will be rotated to the left, to an angle that will 80 be proportional to the thickness of the transmitting materials. The magnitude of the rotation of the plane of polarization is also a function of the wavelength input. This property may be referred to as wavelength 85 dependent optical activity.
(3) It selectively scatters light directed onto the molecular structure. The term scattering is used rather than reflection in order to distinguish from the effect occur- 90 ring on a mirror surface wherein light is reflected at an angle equal to the angle of

[Price 4s. 6d.]

Fig 2 This is the first page of a specification relating to materials which have a cholesteric liquid crystalline phase. Liquid crystals have many important applications in modern electronics (see entry 177). The 'prior art' statement, from line 15 to line 57, gives a very concise summary of the nature of liquid crystals.

Application Numbers
19764—20471

Acceptance Numbers
1400641—1401280

Number 4504
11 June 1975

Official Journal (Patents)

Price 70p

Contents

Fig 3   The *Official Journal (Patents)* has been published weekly by the Patent Office, under various titles, since 1854. It consists mainly of indexes and listings of patent applications showing their stages of progress. A contemporary journalist, thought to be Dickens, referring to it during its early history, called it a 'patentees' newspaper, telling all the current news on the subject'.

A third deterrent to the use of patent specifications is their sheer quantity. Some impression has already been given of the large number of specifications issued each week. It would be foolish to pretend that to search for patents of specific interest is a negligible task. However, all the major patenting countries publish, in addition to the specifications, supporting literature which provides indexes, and often abstracts, so that their patent collections can be searched either for patents issued to named inventors or patenting firms, or for patents relating to given subjects. All patents are classified by subject according to detailed subject classification schemes. Most countries use classification schemes formulated within their own countries but an International Patent Classification (IPC) has now been agreed. Though some countries have adopted the International scheme and use it exclusively, most of the major patenting countries, including the United Kingdom, prefer to persist with their own national schemes but also classify by the International scheme, and print the IPC notations on their published specifications. While it is still generally necessary to make separate searches country by country, using the national classification schemes, systems whereby patents can be searched on an international scale are now being devised and brought into use as various international cooperative ventures get under way.

Overall, the difficulties of searching patent literature are not greater than those of searching the wealth of scientific literature contained in the ever increasing flow of journal articles and reports. However, in the case of the patent specifications a greater degree of responsibility is thrust on the reader in selecting the really valuable contributions. The periodical article is normally vetted by an editorial board drawn from the establishment of the discipline covered by the journal in which it appears. They normally do a good job in protecting the reader from an undue amount of second-rate work but at the same time they may from time to time deny him the opportunity to see accounts of speculative new ideas which are not in line with fashionable theories. What editorial board, for instance, would have had the courage to print Whittle's ideas on jet propulsion at a time when the Royal Air Force, his employers, had so little faith in his proposals that they allowed his patent to lapse. Many inventions which are now universally accepted have passed through a period when only the stubbornness of the inventors, or their backers, have enabled them to survive. There must be many an inventor who at some time would have found himself deeply in sympathy with the mood of Einstein when he wrote in an autograph hunter's notebook the jingle

'a thought that sometimes makes me hazy
am I, or are the others, crazy?'

Farnsworth, whose developments in television were so successful, tells a story of how a professor once gave him four good reasons why his ideas could never work; the original papers published before the breakthrough to successful transistors met with almost complete indifference from most workers in the field. Banting, the discoverer of insulin, who had to fight to be allowed to carry on his research, said later that if he had been familiar in advance with all the complexities of the subject as revealed in the voluminous literature he would never have had the courage to pursue his own work.

The Decca navigation system was twice rejected but won through in the end. It was invented by an American, W. J. O'Brien, in 1937. He approached the United States Navy, the Civil Aeronautics Authority and several United States industrial concerns but could get no support. In 1939 he wrote to a friend at the Decca Gramophone Company in England. Decca persuaded A. V. Alexander, the First Lord of the Admiralty, that the system was of value and it was developed in England, to be used with great success by minesweepers during the landings in Normandy in 1944. After the war a committee headed by Sir Henry Tizard ruled that the system was too limited to be a commercial success in peacetime and official support was dropped. Decca held faith and persisted and were able to demonstrate the value of the system when it was used by colliers during the extreme winter of 1946–47.

Chester F. Carlson, the inventor of the Xerox copier was another who had to hold hard by his faith for a long period. He had proposed the basic concept as early as 1937 and approached no less than twenty industrial concerns in the United States for assistance in developing it. None showed any interest. He persisted and was able to obtain financial support from the Battelle Memorial Institute from 1944 onwards, making such progress that the Haloid Company of Rochester acquired the patent rights in 1950. The Rank Organisation in Britain did not become greatly involved until 1956 when they took over the marketing of the copiers. Their marketing techniques were to some extent novel and may well have been as instrumental in the final phenomenal success as the merits of the invention itself.

Many an inventor has needed to develop new techniques of manufacture and sales in order to get his invention applied. McCormick's reaper was one of the few inventions that was an almost immediate technical success from the start. But while there were no doubts about its capabilities, the farmers of the day just could not afford such a machine and McCormick is credited by some of his biographers as being as great an innovator in the world of business as in that of invention. He introduced a sales system in advance of his time which included fixed pricing, payment by instalments and a servicing guarantee backed by an extensive network of servicing and sales centres.

Since this book aims to draw attention to some of the more significant inventions described in patent specifications the question arises as to how the concept of significance has been interpreted. It might be argued with conviction that all inventions are significant. The concept of novelty embodied in the very definition of invention implies significance, since any step across the threshold of knowledge, however small, is of importance and worthy of record. If despite this an attempt is made to select some of the more worthy inventions, a definition proposed by the Earl of Halsbury provides a useful starting point. Speaking to a meeting of the Institute of Patentees and Inventors in 1971 he said: 'An invention is important or significant if the history of some subject could not be written without reference thereto.'[5] This suggests that a decision should first be made as to what are significant subjects and the histories of these subjects should then be studied with a view to determining the major milestones. This approach has been used to a certain extent in the preparation of this book and has proved very fruitful. In addition an attempt has been made to pick up isolated inventions which might not be related to the development of a major industry but which have had a considerable impact on everyday lives; 'catseyes' are an obvious example of such an invention while surgical gloves represent a less obvious example. In order to trace as many of these as possible, popular textbooks and general accounts of the history of technology have been scanned and the patents sought of any notable inventions that have come to light.

In studying the histories of particular technologies it is often difficult to decide which are the most important inventions. Sometimes a new discovery is made which leads to an explosive leap forward and the associated invention will be accepted by all as being important and significant. More often progress is made in a series of small steps which lead finally to a finished product of some merit. In such cases even the acknowledged experts in the field are hard-pressed to agree which of the small advancing steps are the most significant. The development of colour photography illustrates this well. A study of the textbooks and published histories of this subject shows that there is a measure of agreement as to the importance of some of the early basic ideas, when they were first proposed and who was responsible, but when more recent progress is examined there are so many interconnected developments, each with its own related group of patents, that it becomes very difficult to sort out which have had the greatest effect on the development of processes or products now in common use.

There is a tendency in choosing significant patents to look for those which are, in some way or other, 'firsts'. They may be the first to contain the germ of a new concept, but no idea, however farreaching and basically useful, is of any value until there is a method of applying it to meet a current market demand at a reasonable economic cost. So we have a second concept of a 'first', that of the first to be economically viable. Nobody will nowadays dispute that Glidden's patent for barbed wire fencing is significant. Its practicality in use and reliability in manufacture was such that its applications swept across several continents at a rapid rate. It was by no means the first barbed wire fence, and in fact it was difficult to demonstrate that it contained any basically novel feature. Its value came from the combination of a number of previously disclosed ideas to form an eminently practical end product.[6] The part played by H. Cecil Booth in the story of the vacuum cleaner is commonly accepted as significant, but Booth himself has admitted that none of the principles involved were new, in fact he estimated that there were up to thirty patents for vacuum cleaning

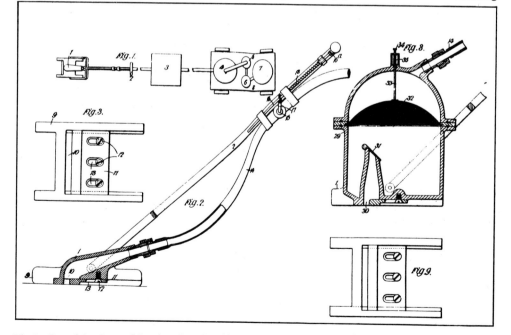

Fig 4   One of the sheets of drawings from Booth's vacuum cleaner patent. Almost the first work done by the original machine was to clean the great blue Coronation carpet under the throne at Westminster Abbey prior to the coronation of King Edward VII. See entry number 349.

devices that predate his.[7] He claims, however, that his model was the first that worked effectively. Mr Justice Farwell, in giving a ruling on the Booth patent said:

'Although utility and novelty are two distinct things, still the fact of utility, when it takes the shape of a great or sudden success for a newly produced article is some evidence of the novelty of the means or process by which so useful an article was produced. If it had not required invention of a sort, it might well be said that the world would long before have discovered what it then found it wanted so much.'[8]

The two cases quoted above are clear examples of significant patents signalled by a sudden and immediate success. In some cases inventions predate the market demand. A classic example of this is that provided by the case of the pneumatic tyre. Thomson invented the first pneumatic tyre with the intention that it should be applied to horsedrawn carriages. It failed dismally, probably because the state of the highways of the time was such that the marginal advantage given by the fitting of the tyres did not warrant the extra expenditure. When Dunlop rediscovered the idea and fitted it to the bicycle it was more quickly in demand. Its introduction was initially resisted by the major manufacturers of the day but they soon gave way in face of an increasingly insistent demand from a public who saw the success that the new tyres were having in racing events. Who is to say that Thomson's once-forgotten patent was less significant than that of Dunlop?

An idea is often formulated before there is a matching technology to bring it to maturity and then the matching of the new technology to the older concept may appear the more significant. For example, Gillette's safety razor was of no value until suitable steels were available for the blades; holography advanced swiftly when the laser was produced to give the necessary coherent light beam, and so on. Most basic inventions need a considerable amount of development effort to produce from them a commercial product. Fleming's discovery of penicillin would have had little impact without the subsequent research on its medical effects and the expenditure that followed to develop reliable methods of quantity production. Perkin's aniline dyes might have been forgotten if he had not expended tremendous energy to find methods of producing sufficient of the necessary chemicals to manufacture his dyes in quantity. The invention of nylon in the Du Pont laboratories, where the first fibres were produced using an improvised spinneret incorporating a syringe needle, also called for a considerable amount of further research to find economical ways of production of the various chemicals involved and to perfect suitable processing machinery.

In the more recent literature on the economics of invention there is a tendency to distinguish between the origin of the basic idea, called the invention, and its development to a commercially viable product or process, called innovation. Often the process of innovation requires a considerable amount of inventive effort and may give rise to its own crop of significant patents. These examples and discussion serve to illustrate that, in the preparation of any list of significant inventions, the decision as to what is, or is not, significant will have to be made subjectively and will vary with the whims and preferences of the compiler. There will be room for controversy as to whether many of the patents in the list warrant their inclusion or should be replaced by others clamouring equally for recognition.

An attempt has been made in this book to include entries for all those well-known inventions for which there are patent specifications. However, not all inventions are patented. There are a variety of reasons for this and some of these are discussed in the next few paragraphs. Most of the basic and fundamental inventions, the round wheel, the use of fire and even more advanced concepts such as glass making, hand methods of paper making and so on, date far back into history, well before man's civilisation became sufficiently sophisticated to see any need for a patent system. We would not therefore expect to find these represented in any list of patents. The first organised system of patents originated in renaissance Italy. The concept was spread through Europe mainly by emigrating Venetian glassblowers whose skill and knowhow were valued and sought after in other countries but who asked for, and were often granted, some guarantee of protection and some form of monopoly rights before they would emigrate and set up in competition with local workers. In England, patents, in the more general sense of 'open letters' (from the Latin *litterae patentes*) had been an established method of granting royal privileges from Norman times. They were used to promote or regulate trade or to provide revenue. Grants of exclusive privileges and guarantees of protection were often given to foreign workers in order to attract new industries from abroad. Some of the merchant and craft guilds were inaugurated under patent grants. It was during the reign of Elizabeth I that the practice of granting patent monopolies became established. Her principal political adviser, William Cecil, Lord Burleigh, was very much aware of the need to encourage English industry and to make the country self-sufficient. The grant of letters patent was a useful instrument towards these ends and many monopolies of inventions or new industries were granted. The Queen's need for money also led her to grant monopolies giving the right to control and regulate established industries. Abuses of these monopolies led to vigorous protests in Parliament so that in 1601 the Queen felt obliged to issue a proclamation revoking the majority of the objectional patents and decreeing fhat any of her subjects who were wronged by the operation of an unfair patent grant could seek redress through the common law courts. The common law courts quickly established that a patent monopoly would only be good in law if it concerned an invention. This was confirmed by Elizabeth's successor, James I, in the *Book of Bounty* which contained the substance of a proclamation issued in 1610, and later in the Statute of Monopolies passed in 1624.

Among the isolated grants of patents for invention granted in England before Elizabeth's time, credit for being the first in usually given to John of Utynam, 'returned of late to England from Flanders at the King's command'. He was granted a twenty-year monopoly in 1449 for a method of making glass which was not previously known in England. John was responsible for supplying glass for the windows of Eton College. The grant to John expressly mentioned a limited period monopoly but did not otherwise greatly differ from earlier protection patents for workers coming from overseas, and its right to be labelled as the first might well be debated. However, it did contain the basic elements of what is expected of a patent today: it was for a process of manufacture and it was an invention in that the process was new to this country. This is still sufficient to define an invention in British law. In most other major patenting countries the invention must be completely novel in that it was not previously known anywhere in the world. This more absolute definition of novelty may become part of the British law in the near future should this be amended to bring it into line with the more general practice and with the proposed regulations for a European patent.

During his period of monopoly John was required to instruct native born Englishmen to perform the process. This was a usual requirement of the protection patents. Later, as the provision of patent specifications became obligatory for the grant of a patent, it gradually became accepted that the publication of the specification was a satisfactory alternative way of ensuring knowledge of the invention was passed on, and the requirement that the patentee should instruct others directly was dropped. This did not altogether relieve the patentee of the responsibility of putting his invention to work. There is provision in today's patent laws to enable any interested party to apply for a compulsory licence in order to make use of a patented invention that could reasonably be put to use more effectively. This provision is designed to prevent the deliberate abuse of a patent monopoly to delay the introduction of a new process or product.

The practice of requiring specifications describing the invention separate from the actual statement of the patent grant was introduced gradually. As patents became more numerous and contests between different claimants more frequent there arose a need for the inventor's claims to be more precisely outlined. At first fuller titles and short descriptions were included with the statement of the patent grant, but since the grant was published before the patent was sealed many inventors objected to revealing their secrets without being sure that a patent would be granted. In some cases a compromise was reached whereby a patent was granted with the proviso that, within six months, the inventor would provide a specification describing his invention. The first patent known to be treated in this way was granted on 3 October 1711 in the name of John Nasmith for a process of preparing and fermenting the wash from sugar and molasses. The practice was regularised from 1734 onwards when all patents granted contained a standard clause invalidating the patent if no specification was lodged within six months. The Patent Law Amendment Act of 1852 changed the position in that it ruled that the period of validity of the patent should start from the date of application rather than from the date of sealing, but the requirement for a separate specification was by this time well established and was retained. So for patents dated after 1734 there will definitely be a specification describing the invention in full, between 1711 and 1734 there may or may not be a separate specification, but for patents entered before 1711 it is unlikely that any useful description of the details of the invention would have been given.

The Patent Law Amendment Act of 1852 also proposed that an Office of the Commissioners of Patents should be set up and should be made responsible both for the running of the patent system and the publication of patent specifications. Bennet Woodcroft became the first Superintendent of Specifications and he immediately set about the task, not only of publishing the newly deposited specifications, but also collecting together earlier specifications and publishing these. Up to that time all letters patent had been recorded on the Patent Rolls. Enrolment could be made at any of three Chancery Offices at the patentees choice; the Enrolment Office, where specifications were entered on the Close Rolls, the Rolls Office or the Petty Bag Office, in both of the latter they were entered on the Specification and Surrender Rolls. From these Rolls Woodcroft extracted details of all patents that had been entered since 1617; where full specifications were available he used these but if none had been entered he used the full text of the grant. All were printed separately in a continuously numbered sequence. Woodcroft chose to begin at 1617 for reasons that were convenient to him at the time, not because there were no earlier patents. It is due to the enthusiasm and energy of Woodcroft that the United Kingdom can now claim to have the longest-running series of printed patent specifications.

A useful annotated list of some of the earlier patents granted by Queen Elizabeth between 1554 and 1600 is provided by Hulme,[9] who also notes one or two worthy inventions for which Elizabeth refused patents. Among these are Harington's water closet and Lee's stocking frame; the latter is described by Hulme as the most original invention of the age. It

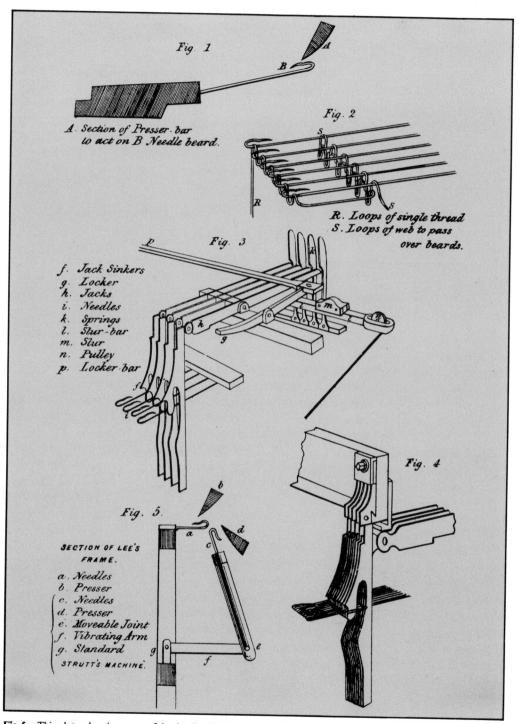

Fig 1

A. Section of Presser-bar
to act on B Needle beard.

Fig. 2

R. Loops of single thread
S. Loops of web to pass
over beards.

Fig. 3

f. Jack Sinkers
g. Locker
h. Jacks
i. Needles
k. Springs
l. Slur-bar
m. Slur
n. Pulley
p. Locker-bar

Fig. 4

Fig. 5.

SECTION OF LEE'S
FRAME.

a. Needles
b. Presser
c. Needles
d. Presser
e. Moveable Joint
f. Vibrating Arm
g. Standard

STRUTT'S MACHINE.

Fig 5   This plate, showing some of the details of Lee's stocking frame, has been copied from William Felkin's book *A History of the Machine-wrought Hosiery and Lace Manufactures* (W. Metcalfe, Cambridge, 1867). Note particularly the bearded needles.

was certainly one of the most intricate pieces of machinery devised up to that time. The reason for the refusal of the patent was linked with a belief that it would deprive many hand-knitters of their means of livelihood, but some historians like to suggest that the Queen's decision may have been unduly influenced by her feminine vanity when she noted with some dismay that the machine-knitted worsted stockings were vastly inferior to the hand-knitted silk hose that she herself wore. Incidentally the story of the Reverend William Lee provides one of the most intriguing tales of motivation in the long saga of the history of invention. There are several versions of this story many of which contradict each other. The following account is taken from Blackner's *History of Nottingham*, published in 1815:

'The inventor of this machine was Wm. Lee, the owner of a freehold estate at Woodborough, the place of his nativity. Deeply smitten with the charms of a captivating young woman of his village, he paid his addresses in an honourable way; but she seemed always more intent on knitting stockings, and instructing pupils in that art than upon the caresses and assiduities of her suitor. He determined therefore to mar her knitting, in order to change her to his views. The former he accomplished in 1589, by the invention of an engine or frame for knitted stockings; a curious and complicated piece of machinery possessing six times the speed of the original mode, and capable of an endless variety of substantial and fancy productions. He gave up the fickle fair one to secure wealth and future fame.'

Many important inventions have been allowed to come into common use without a patent being applied for either because the inventors were unaware of the advantages that a monopoly would give them or because they lacked commercial acumen. An example is the invention of the spinning mule by Samuel Crompton. He perfected the mule when a young man of twenty-six at a time when there was a great need for a method of spinning that would produce consistently high-quality yarn. He allowed his mule to be used by neighbouring spinners on the promise that a subscription would be taken out on his behalf. The use of the mule spread through the industry at a tremendous pace, but both the first subscription list and a later one produced only paltry sums. Crompton was very much discouraged and spent some considerable energy in belated attempts to obtain more fitting recompense, even to the extent of petitioning Parliament.

There have been inventors who have not patented their inventions out of a sense of public-spiritedness. This has often been the case in medical or safety devices. An example is that of Sir Humphry Davy who is usually credited with the invention of the miners' safety lamp, although George Stephenson, the locomotive engineer, may have an equally good claim to this distinction. When strongly urged by one of his friends, John Buddle, to take out a patent so that he might gain financially, Davy replied,—'My good friend, I never thought of doing such a thing: my sole object was to serve the cause of humanity; and if I have succeeded I am amply rewarded in the gratifying reflection of having done so.' David Hughes, the inventor of the loose-contact microphone, is reported as having expressed much the same sentiments, saying he wished his invention to be put to use as soon as possible to enable people to hear properly on the telephone. He had already made his fortune from inventions he had patented earlier.

It is relevant in presenting a list of significant patents to discuss to what extent the publication of a patent specification is a guarantee that the invention claimed in it is new at the time of publication. All systems of patent law require that for a patent to be valid the invention claimed in it must be novel. Definitions of novelty vary from country to country, as we have seen. In most patenting countries there is a stage in the proceedings towards a patent when examiners in the employ of the Patent Office make an examination which includes a search for novelty. The extent of this search also varies from country to country and over time. In the United Kingdom no official search for novelty was made until after 1905. Before then it was presumed that the patentee would make his own search and it was left to the courts to settle disputes as they arose. In 1905 the Patent Office took on extra staff so that official searches could be undertaken before the specifications were accepted and published. However, the official search is limited: the examiner is only obliged to search through British patent specifications over the last fifty years. In practice he will be knowledgeable in the subject field and may quote information from other sources but will not undertake any systematically thorough search through such other sources. In some other countries such as Belgium and Italy there are no formal arrangements for a patent search to be made for novelty and disputes have to be tried through the courts, as for earlier British patents. In yet other countries, such as West Germany and the Netherlands, patent applications are published before any official search is undertaken. There are arrangements for searches to be undertaken after publication if the applicant wishes to proceed further. This examination may lead to changes being made to the specification, which will then be republished. Overall it can be said that the publication of a patent specification is never a guarantee that the invention claimed in it is new, but since any patent may be challenged on grounds of lack of novelty there is a fair certainty that the patentee himself is reasonably convinced that he will be able to defend his claims.

In compiling material for this book the United Kingdom patent has been listed for preference. Foreign patents are often noted either because the United Kingdom patent has been difficult to isolate or, in some cases, because no United Kingdom patent has been filed. This does not, however, mean that the list is limited to inventions originating in this country; a glance through the index of inventors' names will confirm this. The United Kingdom's role in the international world of commerce has been of sufficient importance throughout the history of the patent system to ensure that most inventions of significance would have been the subject of patent applications in this country. The fact that the file of United Kingdom printed patents is the longest running series available is a second inducement for centring on this series. In all historical accounts of invention there is an inevitable tendency for the author to give greater weight to the contributions of his own compatriots. In the history of the typewriter for example, an American would have no hesitation in giving due credit to the work of Christopher Latham Sholes while a French author would feel justified in stressing the importance of the early French work and the Englishman would be at pains to point out that both were predated by the early patent of Henry Mill. In the case of the incandescent lamp, American and English authors might both recognise that significant contributions were made by both Edison and Swan, but the relative weight given to the two contributions might well vary. A similar bias might arise in comparing the telegraph systems of Morse and Wheatstone. Let it be said that the present author is British.

In collecting material it has been necessary to consult a large amount of the extensive literature that exists on the subject of invention. This ranges widely from romanticised simplifications of the origins of well-known inventions, tales of chance discoveries and of individual battles by inventors for recognition, to historical accounts of the farreaching

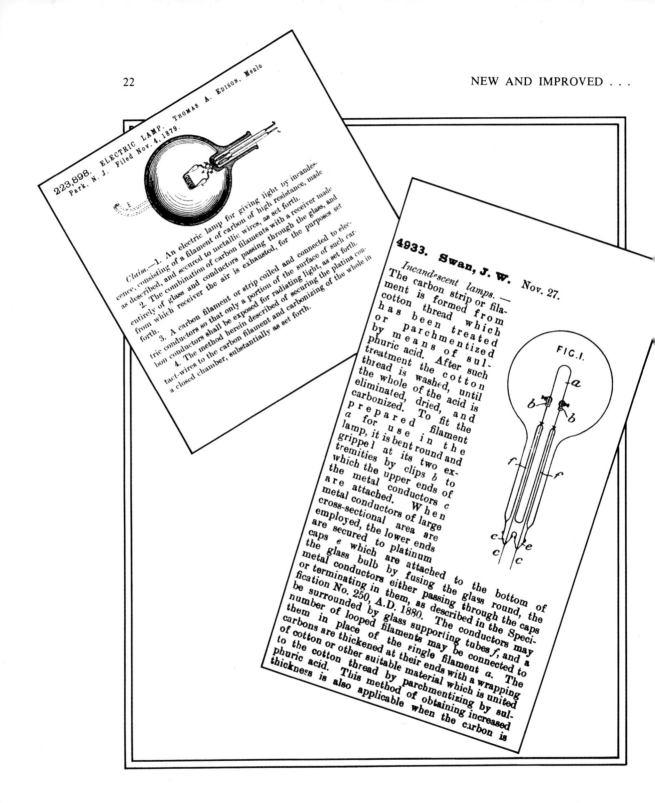

223,898. ELECTRIC LAMP. Thomas A. Edison, Menlo Park, N.J. Filed Nov. 4, 1879.

Claim.—1. An electric lamp for giving light by incandescence, consisting of a filament of carbon of high resistance, made as described, and secured to metallic wires, as set forth.

2. The combination of carbon filaments with a receiver made entirely of glass and conductors passing through the glass, and from which receiver the air is exhausted, for the purposes set forth.

3. A carbon filament or strip coiled and connected to electric conductors so that only a portion of the surface of such carbon conductors shall be exposed for radiating light, as set forth.

4. The method herein described of securing the platina contact-wires to the carbon filament and carbonizing of the whole in a closed chamber, substantially as set forth.

4933. **Swan, J. W.** Nov. 27.

Incandescent lamps. — The carbon strip or filament is formed from cotton thread which has been treated or parchmentized by means of sulphuric acid. After such treatment the cotton thread is washed, until the whole of the acid is eliminated, dried, and carbonized. To fit the prepared filament *a* for use in the lamp, it is bent round and gripped at its two extremities by clips *b* to which the upper ends of the metal conductors *c* are attached. When metal conductors of large cross-sectional area are employed, the lower ends are secured to platinum caps *e* which are attached to the bottom of the glass bulb by fusing the glass round, the metal conductors either passing through the caps or terminating in them, as described in the Specification No. 250, A.D. 1880. The conductors may be surrounded by glass supporting tubes *f*, and a number of looped filaments may be connected to them in place of the single filament *a*. The carbons are thickened at their ends with a wrapping of cotton or other suitable material which is united to the cotton thread by parchmentizing by sulphuric acid. This method of obtaining increased thickness is also applicable when the carbon is

FIG. I.

Fig 6 In England the Edison and Swan business interests combined to produce lamps under the composite name 'Ediswan' to avoid patent squabbles. The item above left is copied from the United States *Official Gazette*. This Gazette has been published weekly since 1872 and has consistently contained illustrated information on current United States patents. Here the claims are reproduced with a drawing, in more recent times an author's summary has replaced the listing of the claims. On the right we see an abridgment for the Swan patent taken from an English abridgment volume.

consequences of the introduction of new technologies, technical and learned accounts of progress in particular sciences and industries, and detailed thoughtful studies into how best to create an environment which will foster invention and encourage the implementation of new ideas.

Most school children are introduced to the story of James Watt and the steam engine by reference to the obvious illustration of the power of steam provided by the domestic kettle when it boils. The repetition of this analogy probably owes more to its usefulness in teaching than to its truthfulness. Watt's contribution to steam engine development was the adoption of the separate condenser. He tells his own *Eureka* story of when he first thought of this:

'It was in the Green of Glasgow. I had gone for a walk on a fine Sabbath afternoon. I had entered the Green by the gate at the foot of Charlotte Street—had passed the old washing house. I was thinking upon the engine . . . and gone as far as Herd's House when the idea came into my head that as steam was an elastic body it would rush into a vacuum, and if a communication was made between the cylinder and an exhausted vessel, it would rush into it and might be condensed without cooling the cylinder. I then saw that I must get quit of the condensed steam and injection water, if I used a jet as in a Newcomen's engine. Two ways of doing this occurred to me . . . I had not walked further than the Golf-house when the whole thing was arranged in my mind.'[10]

There are other stories of sudden revelations associated with the development of the steam engine. Oliver Evans was the first to develop a successful high pressure steam engine in the United States. It is said that he first realised the power of steam when he heard of some youths filling an old musket with water and placing it on a fire where it went off as though filled with gunpowder. There is an account of Savery's discovery of the power of steam in a book by J. T. Desaguliers, published in 1744. Desaguliers suggests that Savery invented the story to substantiate the claim that he derived the principles of his engine independently—

'that having drunk a flask of Florence at a tavern, and thrown the empty flask upon the fire, he called for a bason of water to wash his hands, and perceiving that the little wine left in the flask had filled the glass with steam, he took the flask by the neck, and plunged the mouth of it under the surface of the water in the bason and the water in the bason was immediately driven up into the flask by the pressure of the air.'[11]

There is an attraction about such stories that makes it compulsive for others to repeat and embellish them, and one therefore often doubts their veracity even when the original source may be the inventor himself. One often repeated story that is almost too good to be true concerns Aloys Senefelder, the inventor of lithographic printing. It is said that as a young man he was idly copying out his mother's laundry bill using a wax pencil on a smooth piece of limestone. He accidentally spilt water over the stone and noted that the water wet the stone but not the waxed impression. In this way lithography was born. The story of the invention of the spinning box which paved the way to the development of the synthetic textile industry by showing how to put a twist into rayon fibres is told by the inventor himself. Topham says he came upon the idea when he was out cycling on a wet day and noticed that the mud was thrown out in a circular motion by the wheels of his cycle. Once having decided that the utilisation of centrifugal force would lead to a solution he was quick to produce a practical piece of equipment using very primitive materials. In fact he used 'a flat blacking box about 4 in. diameter and $\frac{5}{8}$ in. deep. I soldered wire lugs to the top and bottom so that I could bind on the lid with string.'[12]

If these stories of flashes of inspiration make the process of invention appear simple we must remember that in each case mentioned above the inspiration came to a receptive and prepared mind. It was Pasteur who said, 'chance only favours intervention for minds which are prepared for discoveries by patient study and persevering efforts'. The discovery of penicillin may well have been due to a chance observation but it needed someone of the calibre of Sir Alexander Fleming to note the significance of the chance event. Tales of chance discoveries are commonplace in the history of invention. Perkin was trying to synthesise quinine, in a home laboratory when on vacation from college, when he discovered the first aniline dye; the ICI chemist who perfected polythene first came across the white waxy solid from which it was derived when attempting an experiment to produce a Diels type of condensation between ethylene and benzaldehyde; Goodyear saw the way towards a method of vulcanising rubber when he dropped a compound of rubber and sulphur on to a hot stove when chatting with friends; Bourdon chanced on the basic principles of his pressure gauge when attempting to correct the faulty work of one of his employees. Bourdon describes how a workman was bending a cylindrical tube into a helical form for use in a still, and clumsily flattened a portion of the tube in the process. In order to restore the tube to its proper shape one end was closed and a force-pump used to force water into the tube. The flattened part of the tube uncoiled itself to a certain extent and Bourdon was quick to see the application for the construction of a pressure gauge.

Even the well organised research team at the Bell Telephone Laboratories who produced the first workable transistors had their moment of chance discovery. The breakthrough to the first practical point contact transistor came as an unexpected result of abortive experiments aimed at finding a field effect amplifier. William Shockley, a principal member of the Bell team, described their methodology of research as one of 'creative failure'—'using failures as opportunities to learn and move ahead'. I doubt whether he looked on the chance discovery of the point contact transistor as a 'failure' but he certainly used the result to move ahead. It was from his detailed theoretical studies to explain the unexpected result that he was led to the invention of the more useful junction transistor.

There have been many discussions of the reasons underlying the frequent recurrence of chance happenings in association with major inventions. John Jewkes said that 'the essential feature of invention is that the path to it is not known beforehand'.[13] This may explain why the large well organised research team with unlimited funds, whose research is channelled to attempting to extrapolate beyond the threshold of knowledge by pushing further forward along well trodden paths, appears to have no greater degree of success in producing the radical basic inventions than the individual worker or small team who, untrammelled by loyalties to existing patterns of thought may, like a comparative novice, 'plunge wildly, wasting much time'. Jewkes goes on to discuss the advantages of the uncommitted mind and lists a number of important inventions which have originated from individual inventors or from relatively small research groups operating outside the mainstream of the industry in which their invention finds its application. In the decade or two following the end of the Second World War the theory that the days of the small inventor were numbered because of the escalating cost of research and the growth of large industrial research teams was much in vogue. In some circles it was believed that no technological problem was insoluble provided sufficient monetary resources and manpower could be allocated to it. Evidence such as that collected by Jewkes has since reduced the credibility of this viewpoint.

It is inevitable that where experimental equipment costs are high and while large research teams grow, an increasing number of potential inventors will find their way into institutionally organised research teams. Such teams will continue to produce a steady stream of 'improvement' inventions, along with a proportion of more radical ideas. But large organisations often tend to stifle radical thought and a reasonable proportion of really significant inventions is still likely to arise from the independent inventor who is able to look at problems from a new point of view. The psychologists call this process 'thinking aside' or 'lateral thinking'. Arthur Koestler develops this theme at some length, drawing an analogy between the act of discovery and the mechanism of humour.[14] Here too it is often the unexpected linking of two completely different concepts that creates an amusing situation, such as the bringing together of the world of the jungle and that of modern technology by the story of the party of witch doctors who arrived in London for a conference in a mumbo-jumbo jet. Koestler describes how Gutenberg conceived his ideas for letterpress printing by a study of the wine press and Kepler worked out his ideas for gravitation by a loosely drawn analogy with the role of the Spirit in the Holy Trinity.

There is no doubt that the sources of invention are numerous and diverse. Those who are concerned to stimulate invention should strive to keep all sources open. There is room for the large research organisation and for the individual inventor, for systematic large-scale empirical work and for careful deductive experiments based on theoretical studies. What is needed above all appears to be an ever ready awareness to see the significance of the unexpected result and to look at problems from fresh viewpoints when it becomes apparent that this might yield more direct results.

NOTES:

1 Derwent Publications Limited.

2 Banks, M. A. L. (Chairman), *The British Patent System. Report of the committee to examine the patent system and patent law*, 1970, HMSO, London. Cmnd. 4407.

3 The Chartered Institute of Patent Agents, Staple Inn Buildings, London, WC1 V7PZ, publish a *List of Patent Agents*, price 50p.

4 Blanco-White, T. A., *Patents for Inventors and the Protection of Industrial Designs*, 4th ed., 1974, Steven and Sons, London.

5 *The Inventor*, Vol. 11, No. 2, June 1971, p. 11.

6 An interesting discussion of this case, and some others illustrating the same point—the incandescent lamp, the telephone, Westinghouse railway brakes and the induction motor—is given by Crotty, F. W., and Dodds, L. B., in *Journal of the Patent Office Society*, Vol. XXX, No. 2, February 1948, p. 18.

7 *Journal of the Newcomen Society*, Vol. XV, 1935, p. 85.

8 *Reports of Patent, Design and Trade Mark Cases*, Vol. XXI, 1904, p. 312.

9 Hulme, E. Wyndham, *The Early History of the English Patent System*, 1909, Little, Brown and Company, Boston.

10 Dickinson, H. W., *James Watt, Craftsman and Engineer*, 1936, University Press, Cambridge.

11 *A Course in Experimental Philosophy*, Vol. II, p. 465.

12 Hard, A., *The Story of Rayon*, 1939, United Trade Press, London.

13 Jewkes, John; Sawers, David and Stillerman, Richard, *The Sources of Invention*, 2nd ed., 1969, Pan Books, London.

Notes on the arrangement
of the List and Indexes

The list of significant patents that follows is arranged in alphabetical sequence under subject headings descriptive of the area of interest in which the patents are deemed to be significant. This general arrangement may be attractive to those who wish to browse through the list, but involves the danger that an invention may be listed under an unexpected heading. To mitigate this an alphabetical subject index is provided which includes both specific and generic headings. The generic headings serve the purpose of linking patents relating to broader topics, such as explosives, motor cars, photography and so on.

A second index lists the names of inventors and patentees in alphabetical order. Many famous and well known names are to be found in this list. Within the body of the main list, the inventor's name, when shown on the patent, is given in capital letters immediately after the patent title. Where the patent has been applied for by an agent or a company to whom the rights have been assigned, the names of these have also been given, but in lower case type. If there is no indication of the individual inventor on the patent then the patentee has been entered in capitals. All names, inventors, agents and assignees, whether personal or corporate, including names mentioned in the notes following the particulars of the patents, are listed in the index.

A third index lists all the patents mentioned in chronological order. If there appears an undue proportion of older patents it should be borne in mind that the method of selection tends to make this inevitable. To a certain extent *significant* has been interpreted in the sense of being *well known*. Even the most outstanding of inventions take some time to become established and recognised. It is easier to judge significance in the light of history. It is a brave man indeed who looks at the multitude of presentday inventions and is prepared to decide which will stand the test of time. See entry under 'STIRLING ENGINE' for an example. Should the gas engine finally prove to be tomorrow's answer to the quest for a noiseless, pollution-free motor car engine, this example will truly demonstrate how the half-forgotten patent of yesteryear could become the significant patent of tomorrow.

Since 1852 all specifications accepted by the Patent Office have been printed in separate leaflet form within a few weeks of acceptance. The patents from 1617 to 1852 which were collected and printed under the direction of Bennet Woodcroft during the years 1853 and 1857 were continuously numbered from one upwards. From 1853 until 1915 the numbers were allocated on an annual basis. In 1916 this practice was brought to an end and patents were allocated numbers in a continuing sequence beginning at number 100,000. So from 1617 up to 1852 and 1916 to date all that is needed to specify a particular patent is its number, but between 1853 and 1915 both a number and a year are required. During the years when patents were numbered annually it is common practice to give number and date separated by an oblique line, e.g. 1784/1911 being patent number 1784 of 1911. This practice has been followed here. The year of publication has also been given for patents in the continuously numbered sequence, with the year in parentheses after the patent number, thus 1,110,791 (1968). In all cases the date given is the date of publication, not the priority date which will invariably be earlier. Where it is apparent that there has been undue delay in publication a comment on the filing date is made in the notes.

Suggestions for Further Reading

One of the purposes of this book is to implant the suggestion in the reader's mind that there is a great deal of useful reading matter in the patent specifications themselves. It is right and proper therefore that we should at this point explain where copies can be obtained or seen.

The major patent holding library in the United Kingdom is the Science Reference Library, now part of the British Library. The Holborn Branch of the Science Reference Library, which is housed in the Patent Office building near Chancery Lane station in London, began life as the Patent Office Library way back in 1852, has always been freely open to the public and now has the wider role of acting as the major reference library for science and technology in the United Kingdom. In addition to a complete set of all British patent specifications this library holds copies of all the other published literature of the Patent Office, together with all the available printed patent literature from foreign countries. All this material is available for reference to any visitor to the library. Photocopies of any of the foreign patent literature will be supplied for the cost of photocopying to visitors to the library or to others applying by post or telex. Copies of any United Kingdom Patent Office publications can be obtained from SALE BRANCH, THE PATENT OFFICE, ORPINGTON, KENT BR5 3RD.

In addition the following public libraries have substantial holdings of patent literature:

Aberdeen
Public Library, Rosemount Viaduct AB9 1GU

Aberystwyth
The National Library of Wales SY23 3BU

Belfast
Central Library, Royal Avenue BT1 1EA

Birmingham
Patents Library, Great Charles Street B3 3JH

Bradford
Yorkshire Central Library, Prince's Way BD5 1NN

Bristol
Central Library, College Green BS1 5TL

Cardiff
Central Library, The Hayes CF1 2QU

Coventry
Reference Library, Bayley Lane CV1 5RG

Edinburgh
Central Library, George IV Bridge EH1 1EG

Glasgow
Commercial Library, Royal Exchange Square G1 3AR

Huddersfield
Central Public Library, Princess Alexandra Walk HD1 2SU

Hull
Central LIbrary, Albion Street HU1 3TF

Leeds
Library of Commerce, Science and Technology, Central Library LS1 3AB

Leicester
Central Reference Library, Bishop Street LE1 6AA

Loughborough
The Library, University of Technology

Liverpool
Patents Library, 5–7 Upper Duke Street, L1 9DU

London
Science Museum Library, South Kensington SW7 5NH

Manchester
Central Library, St Peter's Square M2 5PD

Middlesbrough
Public Library, Victoria Square TS1 2AY

Newcastle upon Tyne
Central Library, New Bridge Street NE99 1MC

Norwich
Central Library, Bethel Street NOR 57E

Nottingham
Central Library, South Sherwood Street
NG1 4DA

Plymouth
Central Library, Drake Circus PL4 8AL

Portsmouth
Central Library, Guildhall Square PO1 2DX

Preston
Harris Public Library, Market Square PR1 2PP

Sheffield
Central Library, Surrey Street S1 1XZ

Swindon
Central Library, Regent Circus

Wolverhampton
Central Library, Snow Hill WV1 3AX

For those who may be prompted by this book to look for further reading on patents and patenting a few selected references are given here.

For accounts of the history of patenting and of the United Kingdom Patent Office the following two books serve as a useful introduction:

Gomme, A. A. *Patents of Invention. Origin and Growth of the Patent System in Britain.* Longmans, Green and Co. for the British Council, London, 1948.
Harding, H. *Patent Office Centenary. A story of 100 years in the life and work of the Patent Office.* London, HMSO, 1953.

A compact account by C. H. Greenstreet of the history of patents throughout the world may be found in the first chapter of Liebesny, Felix (Editor). *Mainly on Patents.* Butterworths, London, 1972. This latter book was compiled to meet the requirements of senior and middle management for information on the working of patent systems both in the United Kingdom and overseas. It also contains information on the protection of trade marks and designs. One of its chapters deals with the subject classification of patents and how searches can be made. A more direct introduction to the searching of British patent specifications is provided by the following free booklet issued by the Patent Office.

Patents, a source of technical information. 1975.

Further free publications issued by the Patent Office with the aim of assisting inventors who wish to apply for patents include:

Applying for a Patent
Information for Patentees
Instructions for the preparation of Specification Drawings.

A very readable account of the wider problems facing the inventor applying for patent protection is given in:

Lees, C. *Patent protection: the inventor and his patent.* London, Business Publications, 1965.

Inventors seeking assistance in the exploitation of their patents may find the following pamphlet of value.

Help for the Inventor obtainable free from the Information Office, National Research Development Corporation, Kingsgate House, 66–74 Victoria Street, London SW1E 6SL.

The standard texts on British Patent Law are:

Patent Law of the United Kingdom. Text, Commentary and Notes on Practice by The Chartered Institute of Patent Agents. Being a revised edition of 'The Patent Acts 1949–1961. Second edition 1968—Annual Supplements to 1974'. 1975, Sweet and Maxwell, London.

Blanco-White, T. A. *Patents for Inventors and the Protection of Industrial Designs.* 4th ed., 1974, Stevens and Sons, London.

A brief outline of the main features of UK patent law and also that of Trademarks, Copyright and Industrial Design is given in:

Blanco-White, T. A. and Jacob, Robin. *Patents, Trademarks, Copyright and Industrial Designs*, 1970, Sweet and Maxwell, London.

A wide-ranging discussion on the history of invention since the nineteenth century and the economic environment most likely to foster inventive effort is contained in:

Jewkes, John; Sawers, David and Stillerman, Richard. *The Sources of Invention.* 2nd ed., 1969, Pan Books, London.

One of the interesting features of this book is that it includes 60 two- or three-page potted histories of particular modern inventions.

Anyone wishing to delve into the psychological literature relating to inventive thought processes will find ample suggestions from the many references given in:

Koestler, Arthur. *The Act of Creation.* 1970, Pan Books, London.

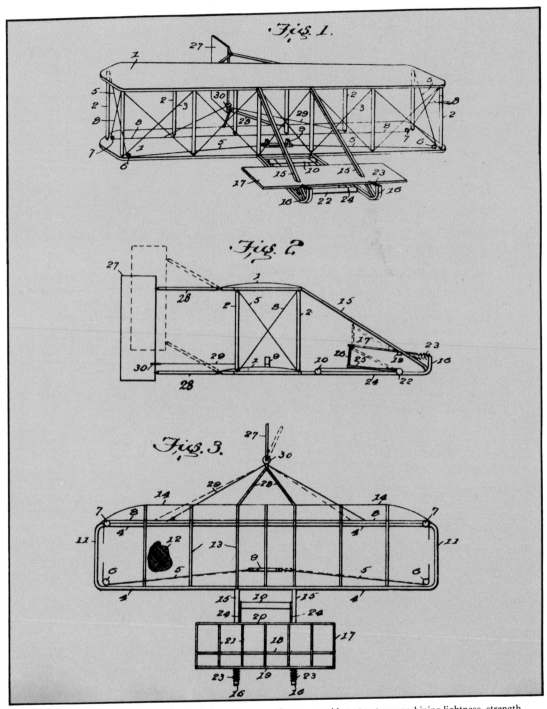

2 The Wrights stated that the objects of their invention were 'first, to provide a structure combining lightness, strength, convenience of construction, and the least possible edge resistance; second, to provide means for maintaining or restoring the equilibrium of the apparatus; and third, to provide efficient means of guiding the machine in both vertical and horizontal directions'.

Alphabetical List of Significant Patents

1 Addressograph

8,945/1896   Improvements in machines for printing addresses on envelopes, wrappers and the like. DUNCAN, JOSEPH SMITH. The relevant United States patents are 558,936 (1896) and 579,706 (1897).

2 Aeroplanes

9,478 (1842)   Certain improvements in locomotive apparatus and machinery for conveying letters, goods, and passengers from place to place through the air, part of which improvements are applicable to locomotive and other machinery to be used on water or on land. HENSON, WILLIAM SAMUEL

9,144/1896   Improvements in flying and soaring machines. PILCHER, PERCY SINCLAIR

6,732/1904   Improvements in aeronautical machines. WRIGHT, WILBUR and WRIGHT, ORVILLE

1,784/1911   Improvements in or relating to flying machines. JUNKER, HUGO

Henson was the first to set out the basic ideas for a powered aeroplane. He failed to recognise the importance of correct shapes for wing cross-sections in order to obtain proper lift. A model of his machine was made by JOHN STRINGFELLOW in 1847–48 and was reputed to be the first powered machine to fly. OTTO LILIENTHAL made studies of birds in flight and was able to apply the results to successful manned gliders, see for example United States 544,816 (1894). His ideas were taken up by Pilcher in England and OCTAVE CHANUTE in the United States, see 13,372/1897, 15,221/1897 and United States 582,718 (1895). Chanute was interested in producing mechanisms that would give automatic stability in flight and was instrumental in transmuting many of Lilienthal's ideas to the Wright brothers. The Wright brothers' patent above refers to the last of their gliders before they turned their attention to powered flight. Junker's patent contains a proposal that all the non-supporting parts of an aircraft should be enclosed in hollow casings designed to minimise air resistance and maximise buoyancy.

3 Aerosol containers

United States 34,894 (1862)   Improved bottle for aerated liquids. LYNDE, J. D.

United States 2,321,023 (1943)   Method of applying parasiticides. GOODHUE, LYLE D. and SULLIVAN, WILLIAM N.

Lynde's patent above is probably the first for an aerosol container; the second, which was the result of work carried out for the United States Department of Agriculture during the Second World War in efforts to find a way of combating the insects which caused disease and discomfort to troops overseas, instigated the widespread development of portable dispensers. Significant developments in between included improvements to the valve mechanisms by G. L. BEBAUR, United States 668,815 (1901) and 711,045 (1902), the first use of carbon dioxide as a propellant, United States 746,866 (1903) R. W. MOORE, and various improvements by E. ROTHEIM, e.g. United States 1,800,156 (1931). Since the war improvements have come thick and fast with new materials being used with varying propellants to produce a range of lightweight containers with widespread uses.

4 Air beds

3,718 (1813)   Making beds, pillows, hammocks, cushions and various other articles of the kind, in a different manner, and of different materials from any hitherto used. CLARK, JOHN

To produce the necessary waterproofing, tick or other suitable material was immersed in a carefully prepared mixture of rubber, turpentine and linseed oil.

5 Air brakes

1,691/1872   Improvements in apparatus for working brakes and communicating signals on railway trains by compressed air, parts of which improvements are applicable generally for retarding locomotives. WESTINGHOUSE, GEORGE, Jnr.

The Westinghouse brake was based on an air pressure system incorporating a valve arrangement for charging, discharging and regulating the pressure in the brake cylinders. There was a compressed air engine which was independent of the movement of the locomotive. The brake was continuous in the sense that it operated on all carriages simultaneously and it was automatic in that it would come into operation should the train part. All these concepts had been patented separately by others before but the combination proposed by Westinghouse was such that it could challenge all other comers at the time, as was shown by its performance in the Newark brake test held in England in June 1875. The brakes were widely used on American railways but there was stubborn resistance to their adoption by some of the major English railway companies. This resistance might have been more quickly overcome had Westinghouse been a less irascible character.

6   Air cushion vehicles (tracked)

913,735 (1962)   Vehicular transport system. LEE, FREDERICK WALTER MADELEY
995,127 (1965)   Improvements relating to vehicles for travelling along a prepared track. COCKERELL, CHRISTOPHER SYDNEY, Hovercraft Development Ltd

For the basic patent for air cushion vehicles see under HOVERCRAFT. Lee's patent above is the first

which discloses a tracked air cushion supported vehicle running at high speed through a specially prepared tunnel with a means for drawing air from ahead of the vehicle and projecting it to the rear. The second patent describes a vehicle running in the open on air cushions along a prepared track.

7   Alabastine

1,620/1882   An improved plaster for casts or mouldings, the coatings of walls, and other analogous purposes. Justice, Philip Middleton for CHURCH, MELVIN BATCHLOR. See also 13,154/1895.

8   Aldrin and Dieldrin

618,432 (1949)   Improvements in or relating to method of forming halogenated organic compounds and the products relating therefrom. HYMAN, JULIUS, Vesicol. The United States patents are 2,635,977 (1953) and 2,676,131 (1954).

9   Alginates

142/1881   Improvements in the manufacture of useful products from seaweeds. STANFORD, EDWARD CHARLES CORTIS

10   Aluminium

5,669–70/1889   Improvements in the production of aluminium. HALL, CHARLES MARTIN

The discovery of an electrolytic, commercially feasible method of producing aluminium was made independently, and almost simultaneously,

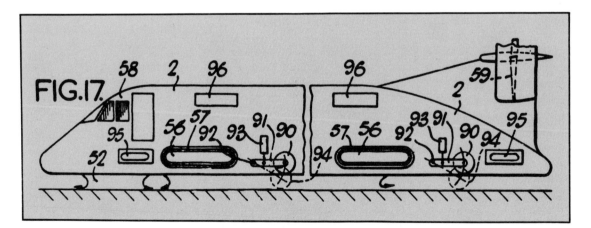

6   This figure is copied from Cockerell's drawings to specification 995,127.

by Hall in the United States and P. HÉROULT in France. The first patents were filed in those countries, United States 400,664–6 (1889) and France 175,711 (1889).

11   Amplifying circuits
Negative feedback   323,823 (1930)
Improvements in or relating to arrangements for amplifying electrical oscillations.
S. G. S. Dickers for N. V. PHILIPS' GLOEILAMPEN FABRIEKEN
Distributed amplifier   460,562 (1937)
Improvements in or relating to thermionic valve circuits. PERCIVAL, WILLIAM SPENCER
Cathode coupled pair   482,740 (1938)
Improvements in or relating to thermionic valve amplifying circuit arrangements. BLUMLEIN, ALAN DOWER

12   Analogue computers
For solving differential equations   745,660 (1956)   Improvements in analogue computers, for solving differential equations. THOMAS, ERIC LLOYD and PAUL, ROBERT JOHN ALEXANDER, Short Brothers and Harland Ltd
For the analysis of physical systems   669,814 (1952)   Improvements in or relating to the electrical analysis of a physical system. WESTINGHOUSE ELECTRIC INTERNATIONAL COMPANY
For the control of power generating plants
871,617 (1961)   Improvements in or relating to computers for power plants. SIEMENS–SCHUCKERTWERKE AKTIENGESELLSCHAFT
See also—884,991 (1961)   Improvements in or relating to electrical generating plant. BROWNLEE, WILLIAM RUSSELL

The above is a selection of some of the more significant of the recent patents in this field. It is difficult to say what might have been the first analogue computer. Sir Robert Watson-Watt suggests that JAMES WATT might be awarded the title of 'father of the analogue computer' for the engine indicator which he devised to monitor the performance of his steam engines. He used these indicators as meters to measure the total amount of power produced in the same way that gas or electricity meters are used. He planned to instigate

a system of charging based on the use of such indicators but met with some violent reaction from the Cornish miners who did not approve of this somewhat unorthodox method of charging and the whole idea was finally dropped. R. Watson-Watt, *Three Steps to Victory*, Odhams Press, 1957, p. 22.

13   Aneroid barometer
10,157 (1844)   Barometers and other pneumatic instruments. Fontainemoreau, Pierre Armand le Comte de, for VIDIE, LUCIEN. The French patent was 12,473 (1844).

This patent provides for thin sheets or diaphragms of metals, glass, indiarubber or other flexible

130   OPTICAL, MATHEMATICAL, AND

A.D. 1844, April 27.—Nº 10,157.

FONTAINEMOREAU, PIERRE ARMAND le Comte de.—(*A communication.*)—" A new mode of constructing barometers and " other pneumatic instruments."

This invention consists " in the application of thin sheets or " diaphragms of metals, glass, india-rubber, or other flexible air-" tight substances," to the above-mentioned instruments, so as to form an elastic cushion susceptible to the slightest variation of pressure of the gas or liquid with which it is contact. The deflexion of the diaphragm is exhibited upon a dial through the intervention of wheelwork.

The drawings show a barometer with the diaphragm corrugated circularly. The diaphragm is supported against the atmospheric pressure by a number of helical springs, and its deflexion is indicated by a hand at the extremity of a screw which is made to rotate by a nut in the diaphragm. A rack and pinion, chain and pulley, or other means may be used to indicate the varying pressures on the diaphragm. Heat and cold are compensated for by a regulating plate of two different metals.

In some instances india-rubber is used in connection with metal for a diaphragm.

[Printed, 8*d*.   Drawing.   See Practical Mechanics' Journal, vol. 2, p. 3.]

13 This is a copy of the abridgment of the patent for the aneroid barometer taken from one of the volumes of abridgments printed under Bennet Woodcroft's direction covering patents issued from 1617 to 1876. Each volume of this series covers patents relating to one of 103 subject classes. The volume from which this is taken is for Class 76, Optical, Mathematical and other Philosophical Instruments. Many of these volumes contain introductory articles sketching the pre-patent history of the subject; for instance, the introduction to this class begins with a reference to the use of the magnetic compass by the Chinese in 2600 BC. Other volumes intersperse the chronological list of patent abridgments with comments on related events of historical interest. To show this, some pages from Class 49, Steam Engines, have been reproduced with entry No. 297 of this book.

airtight substances, to form an elastic cushion or buffer susceptible to the slightest variation of pressure of the atmosphere or fluid with which the diaphragm is in contact. The diaphragm should be very thin and 'corrugated circularly so as to enable it to be depressed or elevated to the greatest extent requisite without rupture'. Provision was made for the use of springs to return the diaphragm to its original shape if its own elasticity were not sufficient. It was obvious that these were not regarded as essential. At the time it was believed that all metal bodies were more or less porous and their elasticity was doubted. so it was only with some courage that Vidie pursued his experiments to the point of success. There was an immediate enthusiastic reception to the invention and the instrument was soon being manufactured in appreciable numbers both in France and in England. Vidie entered into a short business association with EUGENE BOURDON around 1849 for the purpose of increasing the output of his barometers. Bourdon cooperated in their manufacture for a short time but was more interested in manometers to measure steam pressures in the boilers of steam locomotives. This latter association led to considerable argument and litigation in later years.

14   Aniline dye (mauve)
1,984/1856   Dyeing fabrics. PERKIN, WILLIAM HENRY

Perkin was only eighteen years old when he made the experiment that led to his discovery. The experiment was made in a rough home laboratory during an Easter vacation from his studies at the Royal College of Chemistry in London. He was trying to produce quinine from allyltoluidine, by treatment with potassium bichromate. This failed, but it was his curiosity about the red powder which resulted that led him to repeat the experiment using aniline. Manufacture began in June 1857 in a factory at Greenford, Middlesex. Perkin needed to develop safe and economical methods for the production of nitro-benzol and aniline before he could mass-produce his dye.

Some other aniline dye patents are Red 2,153/1866, CARO, H.; and Violet, 1,291/1863, HOFFMAN, A. W.

15   Antiknock fuels (tetraethyl lead)
140,041 (1920)   Improvements relating to fuel for internal combustion engines. KETTERING, CHARLES FRANKLIN and MIDGELEY, THOMAS, Dayton Metal Products Company

16   Antiskid brakes
676,708 (1952)   Automatic braking apparatus for aircraft. TREVASKIS, HENRY WILLIAM, Dunlop Rubber Company Ltd

These brakes, generally known under the name 'Maxaret', were first invented and applied to aircraft landing wheels. The same device is proposed for road vehicles in 847,379 (1960), where it has been reduced to a size which would fit the annular space between the wheel hub and rim of the smaller road vehicle wheels.

17   Artificial silk
2,211/1886   Artificial silk-like filaments from viscous liquids, and apparatus for that purpose. CHARDONNET, HILAIRE DE. French 165,349 (1884)

The story of the invention of rayon began earlier with SIR JOSEPH WILSON SWAN'S patent for nitrocellulose filaments (see under NITRO-CELLULOSE). Swan developed these filaments for electric lamps but did not overlook their textile application. He left some of these filaments at home and the ladies of the household crocheted them into doilies and other articles. It was he who coined the phrase 'artificial silk'. Chardonnet's discovery came when he was experimenting with photographic collodion. He upset a jar containing it and found when it adhered to fingers that it would stretch into bright strands. Chardonnet's filaments were made up into braid by the English firm of Wardle and Davenport, Ltd. It was not until Topham developed his spinning box that it became possible to produce spun yarn.

18   Aspirin (acetyl salicylic acid)
27,088/1898   The manufacture or production of salicylic acid. Newton, Henry Edward for FARBENFABRIKEN VORMALS BAYER & CO., GERMANY

Acetyl salicyclic acid was first prepared by Gerhardt in 1853 and later by Kraut in 1869 by the action of acetyl chloride upon sodium solicylate or salicylic acid. This patent, involving the heating of salicylic acid with acetic anhydride marked the commencement of the commercial development of Aspirin.

In the specification, it is suggested that the substance produced by Kraut was not the *real* acetyl salicylic acid.

19   Atmospheric railway

4,905 (1824)   A method of communication or means of intercourse by which persons may be conveyed, goods transported, or intelligence communicated from one place to another with greater expedition than by means of steam carriages, steam or other vessels, or carriages borne by animals. VALANCE, JOHN

20   Automation of machine tools

715,803 (1954)   Apparatus for the automatic control of machinery. SADLER, ARTHUR LEONARD, Salem Engineering (Canada) Ltd

This apparatus provides for automatic control using cams, templates and so on as used in manual control methods. A development, eliminating the use of cams, templates, etc., and substituting digital data is described in 869,013 (1961), GIDDING AND LEWIS MACHINE TOOL COMPANY. A further development enabling control programmes

for three-dimensional operations to be produced directly from drawings is given in 1,071,531 (1967), FENGLER, WERNER HERMAN. Later improvements include 1,074,368 (1967), DATA RESOLVED TOOLS PTY LTD, which describes a method of checking the shape of a model formed using an automatic programme against an original, 1,080,277 (1967), BENDIX CORPORATION, which allowed for adapting the control system to take account of tool wear and other factors; 1,127,286 (1968), ALLEN, DILLIS VICTOR, IBM, which introduced programs carrying work on pallets from one machine to another; 1,159,799 (1969), BUNKER-RAMO CORPORATION, which suggested the use of a central time-sharing computer controlling many machines; 1,168,104 (1969), GIDDINGS and LEWIS MACHINE TOOL COMPANY, which introduced decision making into the control system; and 1,202,361 (1970), WILLIAMSON, DAVID THEODORE NELSON, Mollins Machine Company Ltd, which showed how a computer could be used to produce a variety of items in relatively short runs.

21   Bailey bridge

553,374 (1943)   Improvements in and relating to the construction of bridges and other metal frame structures. BAILEY, DONALD COLEMAN, Experimental Bridging Establishment, The Barracks, Christchurch, Hampshire

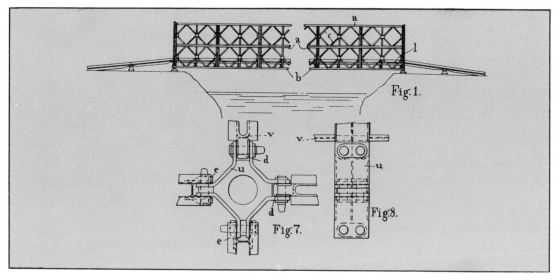

21   A figure from Bailey's specification 553,374.

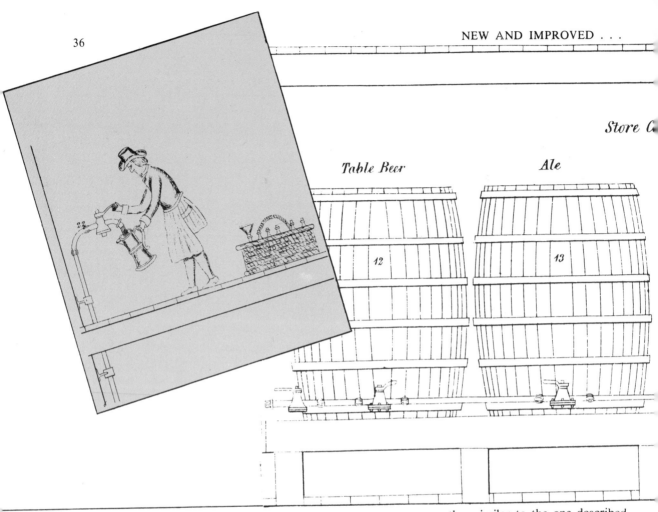

Store C

Table Beer   *Ale*

26 These are two details taken from the drawing submitted by Bramah with specification 2,196. The complete page of drawings measures 20 by 30 inches.

22   Bakelite
1,921/1908   Insoluble condensation products of phenols and formaldehyde. BAEKLAND, LEO

23   Band saw
3,105 (1808)   Machinery for sawing wood, and splitting or paring skins. NEWBERRY, WILLIAM

24   Barbed wire
United States 157,124 (1874)   Wire fences. GLIDDEN, JOSEPH F.

This patent shows the form of barbed wire as we know it today. The idea of a barbed wire fence was not new and many of the earlier versions were in some way or other similar to the one described here. The patent was strongly contested in the United States courts where it was finally concluded that although the difference between the Glidden fence and its contenders were slight, these differences were sufficient to ensure for it the practical and commercial success that evaded the others. The closest of the contenders was United States 74,379 (1868) KELLY, MICHAEL, while the earliest to be patented was that of HUNT, WILLIAM DONISON, filed in the United States as 67,117 (1867).

25   Bassnett sounder
24,916/1893   Improvements in instruments for ascertaining the depth of liquids. BASSNETT, THOMAS

26   Beer pump
2,196/1797   Retaining, clarifying, preserving,

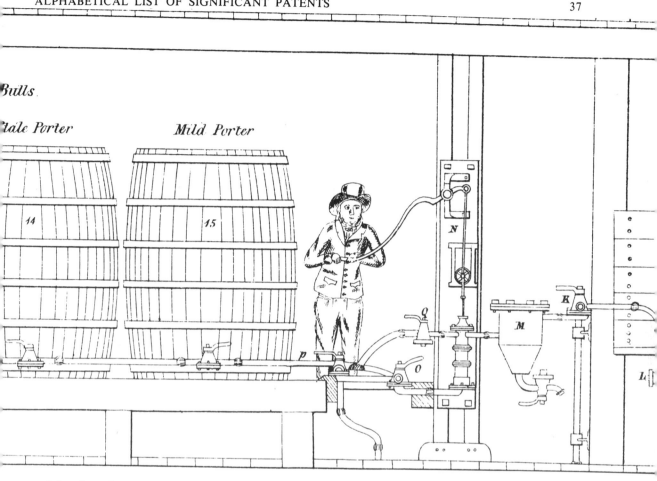

Bulls

ale Porter   *Mild Porter*

14   15

N

Q

M

K

p   *O*   *L*

and drawing off, malt and other liquors; casks and implements for the purpose. BRAMAH, JOSEPH
This patent has a second distinction in that it also describes a hydraulic press, for producing continuous lead or tin pipes by extrusion.

27 Bicycle

Hobbyhorse 4,321 (1818) A machine for the purpose of diminishing the labour and fatigue of persons in walking, and enabling them at the same time to use greater speed, and which he intends to denominate the pedestrian curricle. JOHNSON, DENIS

Ordinary 2,236 (1870) Improvements in the construction of wheels applicable chiefly to velocipedes, and in the driving gear for such vehicles. STARLEY, JAMES and HILLMAN, WILLIAM

The Safety 3,934 (1879) Improvements in the construction of bicycles and other velocipedes, and in apparatus to be used in connection

therewith. LAWSON, HENRY JOHN
Moulton bicycle 907,467 (1962)
Improvements in pedal cycles. MOULTON, ALEXANDER ERIC, Moulton Consultants Ltd

Johnson's hobbyhorse was a direct copy of the 'Draisienne' which had been developed in Paris. The first of the 'Ordinaries', otherwise known as 'boneshakers' or 'penny-farthings' also appeared to originate in France. The Starley patent noted above has been included because it was the first to suggest the use of tensioned spokes for the wheels. Wheel spokes with threaded nipples came a little later, 1,163/1874, GROUT, WILLIAM HENRY JAMES. Lawson's bicycle was not by any means the first of the 'safeties'. Known popularly as 'the Crocodile' it was the first chain-driven bicycle to reach the British public in any quantity. The linked chain was introduced by ALFRED APPLEBY, 8,894/1896. After the general form of the safety had been

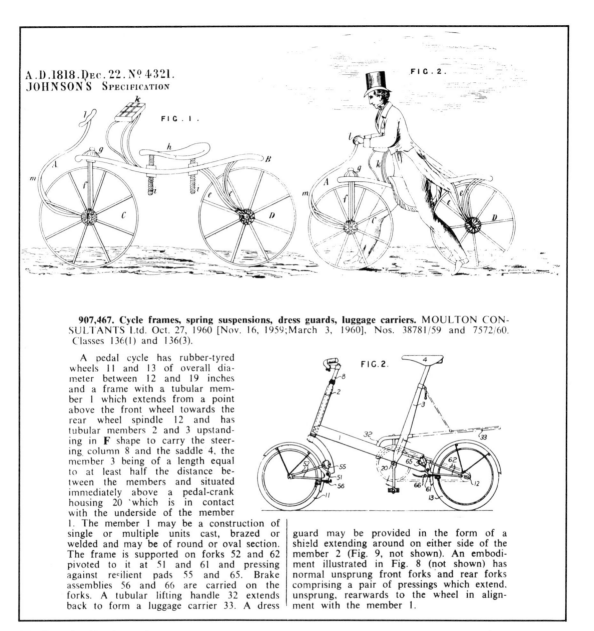

A.D.1818.DEC.22.N°4321.
JOHNSON'S SPECIFICATION

FIG.1.

FIG.2.

907,467. Cycle frames, spring suspensions, dress guards, luggage carriers. MOULTON CON-SULTANTS Ltd. Oct. 27, 1960 [Nov. 16, 1959;March 3, 1960], Nos. 38781/59 and 7572/60. Classes 136(1) and 136(3).

FIG.2.

A pedal cycle has rubber-tyred wheels 11 and 13 of overall diameter between 12 and 19 inches and a frame with a tubular member 1 which extends from a point above the front wheel towards the rear wheel spindle 12 and has tubular members 2 and 3 upstanding in **F** shape to carry the steering column 8 and the saddle 4, the member 3 being of a length equal to at least half the distance between the members and situated immediately above a pedal-crank housing 20 'which is in contact with the underside of the member 1. The member 1 may be a construction of single or multiple units cast, brazed or welded and may be of round or oval section. The frame is supported on forks 52 and 62 pivoted to it at 51 and 61 and pressing against resilient pads 55 and 65. Brake assemblies 56 and 66 are carried on the forks. A tubular lifting handle 32 extends back to form a luggage carrier 33. A dress guard may be provided in the form of a shield extending around on either side of the member 2 (Fig. 9, not shown). An embodiment illustrated in Fig. 8 (not shown) has normal unsprung front forks and rear forks comprising a pair of pressings which extend, unsprung, rearwards to the wheel in alignment with the member 1.

27 From the old to the new in bicycles. The top picture is copied from the original hobbyhorse specification 4,321, while below we have the Patent Office abridgment for the Moulton patent 907,467.

stabilised there was no basic change in bicycle shapes until Moulton introduced smaller wheels.

28 Biro
564,172 (1944)   Writing instrument. BIRO,
LASZLO JOSEF

Biro was awarded a number of patents for various versions of his pen, the first being 498,997 (1938). The specification above relates to a commercially practical version with a capillary constriction near the ball point to aid the maintenance of a continuous supply of ink to the ball which itself was mounted so that its housing would provide an airtight seal when the pen was not in use. For earlier proposals for the use of ball points see

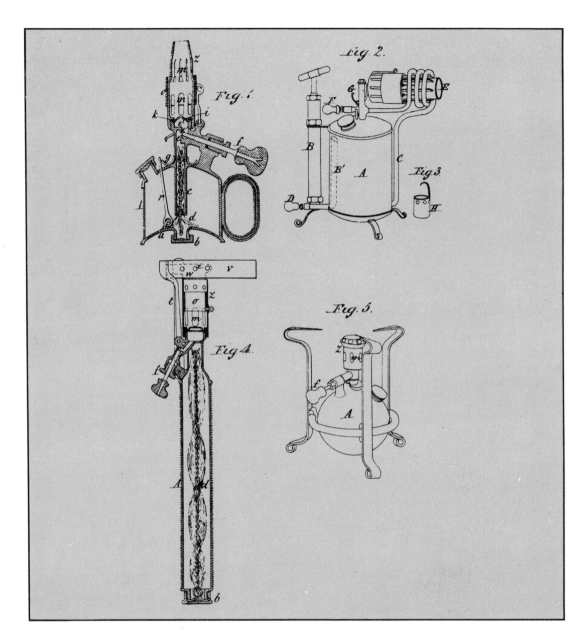

15,630/1888, LOUD, J. J., and 21,747/1891, LAMBERT, E.

29 Blowlamp

8,643/1887 Improvements in lamps for burning volatile oils, suitable for cooking purposes, blowpipes, soldering irons and the like.

SIEVERT, MAX

29 Sievert suggested various uses—with the burner in the horizontal position it could be used as a blow 'pipe' for heating a soldering tool, when used for cooking or laboratory purposes the burner of the lamp, 'instead of having a central opening for the issue of the flame, is formed with an enlarged head having a number of lateral openings through which the flame issues in a number of jets'. This latter configuration gives what we know as the primus stove.

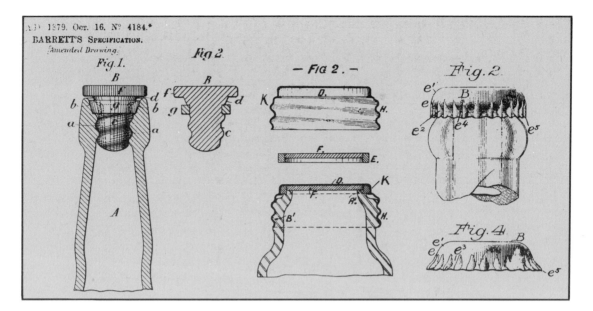

30   Bottle closures
Internal screw stopper   4,184/1879
Improvements in stopping or closing bottles, jars, and similar vessels. BARRETT, HENRY
External screw top with liner   12,629/1889
Improved means of forming and closing the mouths of bottles or jars. RYLANDS, DAN
Crown top   2,031/1892   Improvements in bottle stoppering devices. Lake, H. H. for PAINTER, WILLIAM

31   Bottles
3,434/1887   Improvements in the manufacture of bottles and other hollow ware in glass and in machinery for the same. ASHLEY, HOWARD MATRAVERS

The early bottle makers gathered molten glass by hand, blew the main shape and then finished the neck and mouth of the bottle with hand tools. To make a successful machine the sequence of operations had to be reversed with the neck and mouth of the bottle being moulded first in a ring mould and the main bottle shape being blown after.
This patent, together with two earlier Ashley patents, 8,677/1866 (with ARNALL, J. C.) and 14,727/1886, cover what is known as the blow-blow method in which the main bottle shape is

30   Bottle closures. Internal screw. External screw.
Crown top.

constructed in two stages first in a 'parison' mould and then in a blow mould. All the machines were at first hand-operated. Progress towards automation had to await a method for the automatic gathering and delivery of molten glass. The first successful method was invented in the United States by HOMER BROOKE and applied to the first automatic machines by MICHAEL J. OWENS, Toledo Glass Company, United States 759,742 (1904). The 'gob' feeding method of delivering glass was developed later by KARL E. PEILER, United States 1,199,108 (1916). This used colder and more viscous glass and became the basis for the series of machine produced by the Hartford-Fairmont Companies including the first of the IS machines, United States 1,670,770 (1926), INGLE, HENRY W., Hartford Empire Company, in which, for the first time the molten glass was taken to the bottle making station rather than the other way about. In recent years, since 1930, research has been devoted to producing lighter-weight bottles. This has involved a more scientific study of the shaping of the bottle and methods of strengthening the glass by applying protective coatings, see for instance, United States 2,831,780 (1955), DEYRUP, A. J. and 3,161,534 (1961), DETTRE, R. H.

32 Bourdon pressure gauge

12,889 (1849) Instrument for measuring, indicating and regulating the pressure and temperature of air, steam, or other fluids, and for obtaining from them motive power. COWPER, CHARLES. The original French patent was 4,408 (1849) in the name of EUGENE BOURDON.

Pressure gauges were brought into use on German railways at about the same time, see for example Prussian patent 3 (1849), RABSKOPF, J. C. Bourdon's instrument, which contained the original conception of the oval tube element, was a simpler instrument.

33 Bowden cable

25,325/1896 New and improved mechanism for transmission of power. BOWDEN, ERNEST MANNINGTON

The first of the many applications of the Bowden cable was to the brakes of bicycles, see 14,402/1897 and 1,196/1898, both also due to Bowden. The cantilever action of the modern bicycle cable brake came much later, 330,478 (1930) Meredith, Léon for JEAY, MARCEL.

33 The Bowden cable is one of a number of inventions developed mainly because of their applications to the bicycle but which had much wider uses. The early evolution of the ball bearing, the pneumatic tyre and tubular construction all of which were later of significance in the development of the car and aeroplane industries all owed much to the demands of the bicycle.

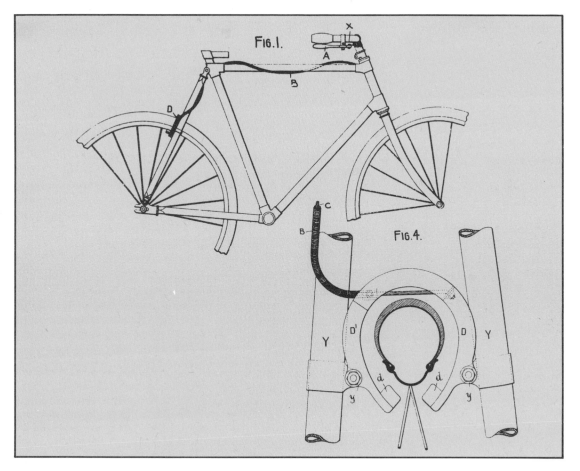

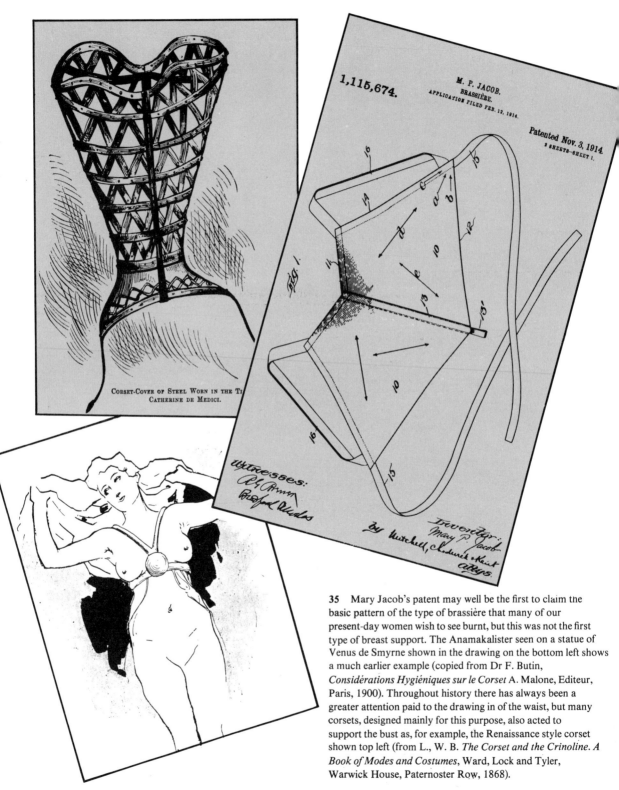

CORSET-COVER OF STEEL WORN IN THE TI
CATHERINE DE MEDICI.

1,115,674.

M. P. JACOB.
BRASSIÈRE.
APPLICATION FILED FEB. 12, 1914.

Patented Nov. 3, 1914.
2 SHEETS—SHEET 1.

35 Mary Jacob's patent may well be the first to claim the basic pattern of the type of brassière that many of our present-day women wish to see burnt, but this was not the first type of breast support. The Anamakalister seen on a statue of Venus de Smyrne shown in the drawing on the bottom left shows a much earlier example (copied from Dr F. Butin, *Considérations Hygiéniques sur le Corset* A. Malone, Editeur, Paris, 1900). Throughout history there has always been a greater attention paid to the drawing in of the waist, but many corsets, designed mainly for this purpose, also acted to support the bust as, for example, the Renaissance style corset shown top left (from L., W. B. *The Corset and the Crinoline. A Book of Modes and Costumes*, Ward, Lock and Tyler, Warwick House, Paternoster Row, 1868).

34 Bowls
687/1872   Turning bowling green bowls.
TAYLOR, THOMAS

This was the first machine for shaping bowls and giving them the specific degree of bias required. Taylor introduced the system of describing the various degrees of bias by numbers 1 to 5, instead of using vague terms such as narrow, medium and wide. These number standards are now used on an international basis although intermediate degrees have been introduced. He constructed a testing table in his factory on which balls could be rolled and their bias checked in controlled conditions.

35 Brassière
United States 1,115,674 (1914)   Brassière.
JACOB, MARY P.

36 Brightening agents
442,530 (1935)   Improvements in the preparation of safety paper. IMPERIAL CHEMICAL INDUSTRIES, LTD

This patent concerns a method of impregnating paper so that it would glow in ultraviolet light. Any attempt made to bleach out writing on the paper could subsequently be detected. The use of brightening agents in detergents came later, see for instance 584,484 (1947), THOMAS, RICHARD, Lever Brothers and Unilever Ltd.

37 Calculating machines
United States 159,244 (1875)   Calculating machines. BALDWIN, FRANK S.
United States 209,416 (1878)   Calculating machines. ODHNER, WILLGODT, Konisberger and Co.
United States 388,116–388,119 (1888) Calculating machines. BURROUGHS, WILLIAM S., American Arithmometer Company
17,138/1888   An improved adding machine.
FELT, DORR EUGENE and TARRANT, ROBERT

The machines covered by the first two patents above were the first to incorporate the flat pin-wheel in place of the stepped cylinder of earlier machines. The Baldwin machine could be used for addition, subtraction, complex multiplication, division and the extraction of square and cube roots. The results were preserved by a ribbon printing device or by perforations on paper. Odhner's patent stated that the calculations could be performed without any labour other than that required to set and rotate certain numbered counting wheels and to adjust the slide carrying a series of recording wheels. The series of Burroughs's patents showed a machine in which the numbers were set on the registers by depressing keys and both the numbers entered and the results of the calculations were printed on paper. Felt and Tarrant's machine also incorporated a key mechanism. The next major improvement was made by OTTO STEIGER who made use of the Bollée tongued plate to produce a machine that multiplied directly instead of by multiple addition.

38 Camera
6,950 (1888)   Improvements in or relating to cameras. Boult, Alfred Julius for EASTMAN, GEORGE

This was the first 'Kodak' mass-produced camera. 'Pull the string, turn the key, press the button' so

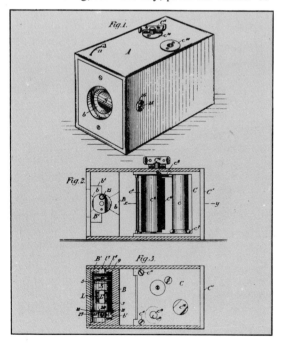

38 The 'snap' camera.

went the advertising. This box, roll-film camera brought photography to the mass market. It was in fact the culmination of a series of improvements to the earlier glass-plate cameras. The first suggestion for applying roll film to a camera appears to have been made by LEON WARNERKE, the English patent was 1,139/1854, MELHUISH, ARTHUR JAMES and SPENCER, JOSEPH BLAKEY. However, there is no evidence of this roll holder coming into practical use and Eastman and WILLIAM H. WALKER independently produced a roll holding apparatus around 1884. The first photographic film was called 'stripping' film, United States 306,594 (1884), EASTMAN, GEORGE, since the gelatinous negative film was attached to a paper roll and had to be stripped off on to a glass plate for printing purposes. Transparent films were developed later by a research worker of 'Kodak', said to be the first chemist to be employed by industry specifically to do research, the patent is United States 417,202 (1889), REICHENBACH, HENRY M., Eastman Dry Plate and Film Company:

39   Camera focusing
533,876 (1941)   Improvements relating to automatic focusing devices for cameras. HUITT, LIONEL HUBERT

This relates to a photo-electric method of automatic focusing.

40   Carbon paper
3,214 (1809)   Pen to promote facility in writing: black writing ink, the durability of which will not be affected by time or change of climate. FOLSCH, FREDERICK BARTHOLOMEW and HOWARD, WILLIAM

The coating for the carbon paper was to be made of Frankfort black mixed with melted, saltfree butter. This was smeared over paper with a sponge, allowed to dry, then pressed flat before use.

41   Carbon fibre
1,110,791 (1968)   The production of carbon fibres. JOHNSON, WILLIAM; PHILLIPS, LESLIE NATHAN and WATT, WILLIAM, National Research Development Corporation

See also 911,542 (1962)   TOKAI DENKYOKU SEIZO KABUSHIKI KAISHA, which contains the germ of the basic concept.

42   Carborundum
17,911/1892   Carbonaceous compounds. ACHESON, E. G.

Acheson coined the name CARBORUNDUM because he thought, erroneously, that the abrasive material he had made was composed of CARBON and CORUNDUM; it is in fact a compound of carbon and silicon. He described the experiment which led to its discovery as follows—'An iron bowl, such as plumbers use for holding their melted solder, was attached to one lead from a dynamo and filled with a mixture of clay and powdered coke, the end of an arc light carbon attached to the other lead was inserted into the mixture. The percentage of coke was high enough to carry a current, and a good strong one was passed through the mixture between the lamp carbon and bowl until the clay in the centre was melted and heated to a very high temperature. When cold, the mass was examined. It did not fill my expectations, but I by sheer chance, happened to notice a few bright specks on the end of the arc carbon that had been in the mixture.' One of these specks, when mounted on the end of a lead pencil and drawn across a pane of glass, cut it like a diamond.

43   Carburettor
16,072/1893   Producing explosive mixtures in hydrocarbon engines. MAYBACH, WILHELM

44   Carding machines
628 (1748)   Machine for carding wool and cotton by hand or water. BOURNE, DANIEL
636 (1748)   Machine for carding wool, cotton and raw silk. PAUL, LEWIS

These were the earliest carding machine patents.

45   Carpet sweeper
3,399 (1811)   A sweeping machine or brush or improvements on a sweeping machine or brush, or sweeping machines or brushes. HUME, JAMES

The machine described in this specification

1,110,791. Carbon fibres. NATIONAL RESEARCH DEVELOPMENT CORPORATION. 26 April, 1965 [24 April, 1964; 29 Dec., 1964], Nos. 17128/64 and 52708/64. Heading C1A. [Also in Division C3]

Carbon fibres are prepared from organic polymer fibres, the fibres having been submitted at some stage of the conversion to the combined effect of heat and tension. The polymer fibres may be exposed to an oxidizing atmosphere at a temperature of 200—300° C. for sufficient time for substantially complete oxygen permeation of the fibres while under a tension such that there is little or no longitudinal shrinkage, and then carbonized by heating to about 1000° C. under non-oxidizing conditions, optionally while still under tension. Preferably the carbon fibres are then heated to 2000° C. or more under non-oxidizing conditions. In the examples the starting material is multi-filament polyacrylonitrile yarn. Carbon fibres may be made from polyacrylonitrile fibres without a preliminary oxidizing step by submitting the polyacrylonitrile fibres to the combined effect of heat and tension at a carbonizing temperature of about 1000° C. under non-oxidizing conditions, optionally followed by further heating to at least 2000° C. The carbon fibres of the invention may have tensile strengths of from 120×10^3 to 260×10^3 p.s.i. and a Young's modulus of from 20×10^6 to 60×10^6 p.s.i. The carbon fibres may be used as reinforcement in synthetic plastic materials (see Division C3).

1,110,791. Polymer compositions containing carbon fibres. NATIONAL RESEARCH DEVELOPMENT CORPORATION. 26 April, 1965 [24 April, 1964; 29 Dec., 1964], Nos. 17128/64 and 52708/64. Headings C3B, C3P and C3R. [Also in Division C1]

Carbon fibres which are prepared from organic polymer fibres by a process which includes submitting the fibres at some stage of the conversion to the combined effect of heat and tension (see Division C1) are used as reinforcing elements in composite materials. In Example 3 the following constituents were compounded in a rubber mill:—a fluoroelastomer copolymer of vinylidene fluoride and hexafluoropropylene, MgO, dicinnamylidone hexamethylene diamine (curing agent) and the carbon fibres of the invention. In Example 4 a mixture of polyester resin, methyl ethyl ketone peroxide and a 6% solution of cobalt naphthenate in white spirit was poured over a bundle of carbon fibres and the resin cured and set at ambient temperature. In Example 5 the procedure of Example 4 was repeated with an epoxy resin and polyamide hardener. In Example 6 the carbon fibres were impregnated with a resin comprising diphenyl oxide and paradichloroxylylene in a dichloroethane solution. After evaporation of the dichloroethane the fibrous mass was shaped and heat-cured at 180° C. and 500 p.s.i. for 2 hours.

41 The Patent Officer examiners are organised in small groups, each of which is responsible for the examination of patents within a strictly limited subject field. Applications for patents will be sent, after a preliminary classification, to the group within whose subject area it falls. If the subject matter bridges the scope of more than one group the application will be passed between the groups concerned. Each group is responsible for maintaining a file of the patent specifications relating to their subject. This file includes a complete copy of each specification and a summary. This summary, known as an abridgment, is written by one of the group's examiners and, if the application is accepted, will later be published by the Patent Office as abridgments. The abridgments are published weekly in looseleaf form and in bound volumes at intervals of about 7 months (every 20,000 specifications). On publication they are grouped under subject divisions and supplemented with detailed subject indexes thereby providing the searcher with a useful aid in searching specifications by subject. Both the abridgments above relate to the same patent for carbon fibres, note how each has been drafted to meet the particular needs of the individual examining group.

consists of a revolving brush contained in a box on castors and fitted with ledges to catch the dust swept up. It is very similar to some in use today.

46   Cathode ray tube receiver

27,570/1907   New or improved method of electrically transmitting to a distance real optical images and apparatus therefor. ROSING, BORIS

47   Catseyes

457,536 (1936)   Improvements related to blocks for road surface markings. SHAW, PETER

This was the first 'catseyes' road marker to clean itself as traffic ran over it. Such markers have been in common use since. For an earlier catseye patent see 436,290 (1935).

48   Cellophane

3,929/1912   Cellulose films. BRANDENBERGER, JACQUES EDWIN

Brandenberger was a French dye chemist and had been experimenting to produce a wrapping material with the easily cleanable properties of cellophane since 1900. The material described in the patent was not waterproof. The United States rights were taken over by DU PONT who put in hand further research and came up with a waterproof product, 283,109 (1928), CHURCH, W. H. and PRINDLE, K. E., Du Pont Cellophane Co., Inc.

47   Figures from specification 457,536.

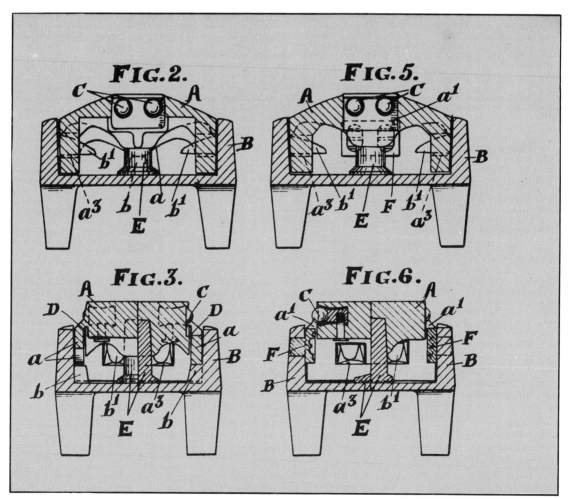

49 Celluloid
1,025/1871 Plate or base for artificial teeth.
LAKE, WILLIAM ROBERT
3,101/1872 Compounds of pryoxylene or gun cotton. LAKE, WILLIAM ROBERT

The inventor of celluloid was the American JOHN WESLEY HYATT who, with his brother ISAIAH, took out United States patents 91,341 (1869) and 105,338 (1870). He set to work on the invention in an effort to win a prize of $10,000 offered by a New York firm for anyone who could find a practical substitute for ivory in the manufacture of billiard balls. He made a number of useful plastic compositions before turning his attention to experiments with nitrocellulose. He had problems with some of his first nitrocellulose materials. A lighted cigar applied to the ball at once resulted in a serious flame and occasionally 'the violent contact of the balls would produce a mild explosion like a percussion guncap'. One billiard saloon proprietor in Colorado wrote to Hyatt to say that he did not mind these noises himself but it was a trifle dangerous, for every man in his saloon immediately pulled a gun.

50 Centrifuge
4,459/1878 Apparatus for separating fluids of different specific gravities. LAVAL, GUSTAV DE

51 Cereal flakes
13,094/1915 Improvements in puffed cereal flakes and method of preparing same. MARTIN, FRANK, Potsum Cereal Co., Ltd

52 Chirp radar
604,429 (1948) Improvements in and relating to systems operating by means of wavetrains.
SPROULE, DONALD ORR; HUGHES, ARTHUR JOSEPH, Henry Hughes and Son Ltd

53 Chlorine manufacture
1,403/1868 Manufacture of chlorine. DEACON, HENRY
2,476/1870 Improvements in the manufacture of bleaching powder by the use of chlorine when diluted with inert gases. DEACON, HENRY

54 Chlorozone
2,349/1878 Liquids for decolorising and bleaching. CLARK, ALEXANDER MELVILLE

55 Chocolate
514 (1730) For the expeditious, fine, and clean making of chocolate by an engine. CHURCHMAN, WALTER

The rights to this patent were brought up by Dr Joseph Fry and formed the base upon which Fry's chocolate firm was developed.

56 Chromium plating
243,046 (1925) Improvements in and connected with the production of solutions containing oxides of chromium, and with the electrolytic separation of chromium from such solutions. LIEBREICH, ERIX. The German patent was 448,526 (1924).

The first mention of the electrodeposition of chromium was made by ROBERT WILHELM BUNSEN in 1854 but it was many years before practical methods were developed. Dr Liebreich's method was developed by him in Berlin and was the first to lead to commercial applications.

57 Chronometer
1,328 (1782) Escapement and balance, to compensate the effects of heat and cold in pocket chronometers or watches, also for incurvating the two ends of the helical spring, to render the expansion and contraction of the spring concentric with the centre of balance. ARNOLD, JOHN. See also 1,113 (1775) and 1,354 (1783).

JOHN HARRISON is often given credit as the inventor of the chronometer. However, his methods of construction were tedious and costly, he made very few and did not patent.

58 Chubb lock
4,219 (1818) Certain improvements in the construction of locks. CHUBB, JEREMIAH

In 1818 a serious robbery took place in Portsmouth Dockyard and the government of the day offered a reward of £100 for a lock which could not be picked. Chubb was in the hardware

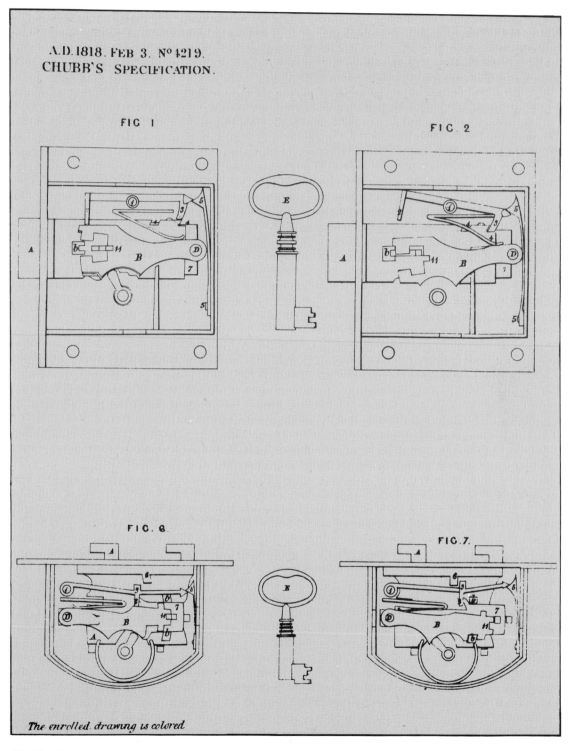

A.D. 1818. FEB 3. Nº 4219.
CHUBB'S SPECIFICATION.

FIG 1

FIG. 2

FIG. 6.

FIG. 7.

The enrolled drawing is colored

58   The Chubb lock.

business in the port and promptly' secured the reward, thereafter devoting himself almost exclusively to the manufacture of locks and safes.

59   Cigarette making machine
14,881/1895   Cigarette making machines. BARON, B.

60   Cigarette packing machine
227,934 (1925)   Improvements in automatic machines for packing cigarettes and articles of similar shape. MOLLINS, WALTER EVERETT

61   Cinecamera
10,131/1889   Improved apparatus for taking photographs in rapid series. GREENE, WILLIAM FRIESE and EVANS, MORTIMER

This is the first cinecamera in which the successive pictures were taken on a strip of film with a single camera. An earlier device involving the use of several camera lens systems to take sequential pictures spaced out on a single photographic plate was described in 423/1888, LE PRINCE, LOUIS AIMÉ.

62   Cinerama
518,905 (1940)   Improvements in motion picture theatres and methods of projecting pictures and producing sound effects therein. WALLER, FRED and WALKER, RALPH

First called Viturama, this system needed considerable expenditure on development before being first tried out on a commercial basis in the 1950s.

63   Clutch, friction
263/1863   A new or improved coupling and break for transmitting or regulating or arresting motion. WESTON, THOMAS ALDRIDGE

64   Coalite
14,365/1906   Improvements in or relating to fuel. PARKER, T.

65   Cold rolling of metals
1,398 (1783)   Hardening and stiffening copper, and reducing same from large masses to any diameter from any length and form, by the use of grooved rollers: also hardening and stiffening brass, iron, steel, mixed and compound metals that will bear drawing or beating out by forge or hammer in either a hot, warm or cold state. WESTWOOD, JOHN

66   Colour photography
Early patent for three-colour process
2,973/1876   Improvements in photography in colours, and in apparatus for that purpose. Morgan-Brown, William for DUHAURON, LOUIS DUCOS
First monopack   Austrian 42,478 (1910) Verfahren zur Herstellung farbiger Photographien. SHINZEL, KARL
Origins of Agfacolor   15,055/1912 Improvements in or relating to the production of photographs in natural colours. FISCHER, RUDOLF 481,501 (1938)   Improvements in the production of colour photographs. Groves, Wilfred William for I.G. FARBENINDUSTRIE AKTIENGESELLSCHAFT
Origins of Kodachrome   United States 1,516,824 (1924)   Colour photography. MANNES, LEOPOLD D. and GODOWSKY, LEOPOLD. See also United States 1,659,148 (1928).
United States reissue 18,680 (1932)   Colour photography. TROLAND, LEONARD THOMPSON. The British patents are 370,908 (1932) and 382,320 (1932).

67   Colour television
System which permitted the representation of colour information and black and white information by different signals   524,443 (1940)   Improvements in or relating to television systems. VALENSI, GEORGES
Colour dot-sequential system   676,670 (1952) Improvements in colour television apparatus. RADIO CORPORATION OF AMERICA
Constant luminance principle   689,356 (1953) Color-television system. HAZELTINE CORPORATION. See also 689,821 (1953).
PAL system   1,005,855 (1965)   Improvements in or relating to colour television receivers. TELEFUNKEN PATENT VERWERTUNGS-GmbH

A.D. 1876, *22nd July.* N° 2973.

Photography in Colours.

LETTERS PATENT to William Morgan-Brown, of the Firm of Brandon and Morgan-Brown, Engineers and Patent Agents, of 38, Southampton Buildings, London, and 13, Rue Gaillon, Paris, for the Invention of " IMPROVEMENTS IN PHOTOGRAPHY IN COLOURS, AND IN THE APPARATUS FOR THAT PURPOSE." A communication from abroad by Louis Ducos Duhauron, of Agen (Lot and Garonne), France, Chemist.

Sealed the 17th November 1876, and dated the 22nd July 1876.

COMPLETE SPECIFICATION filed by the said William Morgan-Brown at the Office of the Commissioners of Patents on the 22nd July 1876.

WILLIAM MORGAN-BROWN, of the Firm of Brandon and Morgan-Brown,
5 Engineers and Patent Agents, of 38, Southampton Buildings, London, and 13, Rue Gaillon, Paris. " IMPROVEMENTS IN PHOTOGRAPHY IN COLOURS, AND IN THE APPARATUS FOR THAT PURPOSE." A communication from abroad by Louis Ducos Duhauron, of Agen (Lot and Garonne), France, Chemist.

GENERAL DESCRIPTION OF THE SYSTEM.

10   The problem of photography in colors such as I express it does not consist in submitting to the luminous action a surface prepared in a manner to appropriate and preserve at each point the colouration of the rays which strike it. What I ask is that the sun avails itself judiciously of an invariable palette, to which it is trusted. It must be that he only charged with the choosing and the mixing of the
15 colours furnished to him will obtain incomparable copyist the most learned and the most authentic of the representations of nature. The problem being defined in these terms, an observation that I have checked by numerous experience has served for the point of departure to resolve it. That observation is this : In opposition to the colors of the spectrum, the scale of which is formed of innumerable shades, and
20 allows of our distinguishing and naming seven principal ones, the coloring substances which serve to express them are reduced to three; red, blue, yellow. This triplicity of colors has been of long date experimentally but confusedly recognised ; if it had been verified beyond possibility of doubt it would have long since been an axiom,

[*Price 8d.*]

66   This is the first page of the specification in which Duhauron outlines the basic three-colour process on which colour photography, colour printing and colour television have been built.

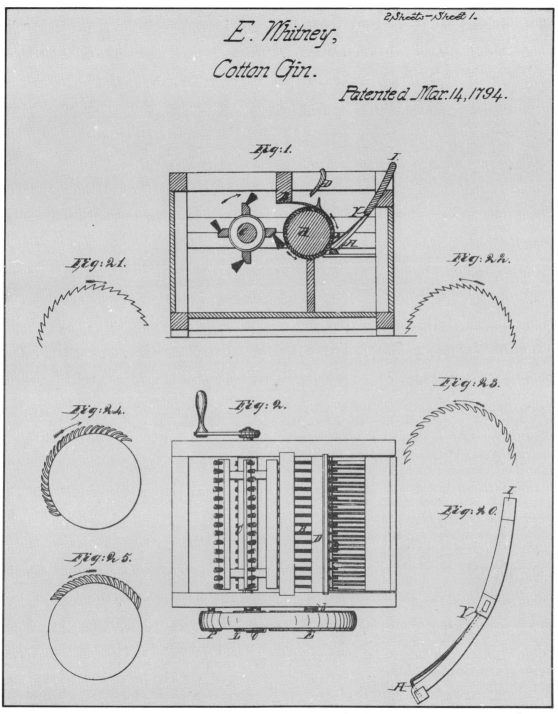

73 Whitney's cotton gin must rate as one of the all-time greats amongst American inventions. Some accounts suggest that Whitney devised the basic principle underlying the effectiveness of the device by analogy with the rotary motion of the rough upper surface of a cow's tongue when licking corn stalks.

JAMES LOGIE BAIRD gave the first public demonstration of colour television as early as 1928. It was the PAL system that was used when colour television broadcasting was introduced in the United Kingdom in 1967. The basis of the system was contained in an earlier patent 702,182 (1954) HAZELTINE CORPORATION

68 Concertina
5,803/1829   Wind musical instruments.
WHEATSTONE, CHARLES

69 Continuous casting
United States 2,135,183–4 (1938)   Process for continuous casting of metal rods. JUNGHANS, SIEGFRIED
677,023 (1952)   Process and installation for the continuous casting of materials of high melting point such as steel. JUNGHANS, SIEGFRIED

The Rossi-Junghans method of continuous casting for non-ferrous metals was invented in 1927. The pair of patents referred to in the first entry above describe the basis of the method. At first the method could only be applied to metals with relatively low melting temperatures. The second entry above refers to a patent which discloses improvements which permit the method to be applied to steel. Examples of more recent developments are 1,184,837 (1970), MITSUBISHI JUKOGYO KABUSHIKI KAISHA, which shows how a vessel of circular cross-section can be made by progressive casting, from a mould of a continuous helical strip of metal, fresh molten metal being deposited on previously solidified metal, and 1,157,977 (1969) METTALURGICHESKY ZAVOD IMENI V. I. LENINA, which describes the first case in which the metal is held and shaped on a descending support in an electromagnetic field.

70 Continuous hot strip mill
256,798 (1926)   Improvements in metal rolling.
TYTUS, JOHN BUTLER

There had been earlier attempts at continuous rolling which had not reached beyond the experimental stage. The first mill to use this method came into operation in 1932. A similar

process developed by H. M. NAUGLE and J. M. TOWNSEND of the COLUMBIA STEEL COMPANY but for rolling long sheets, as opposed to the short wide sheets of the Tytus method, was put into operation in 1926.

71 Continuous rod mill
1,935 and 2,520/1862   Rolling wire and other rods or bars of metal. BEDSON, GEORGE

72 Cordite
5,614 and 11,664/1889   An improvement in the manufacture of explosives and ammunition. ABEL, FREDERICK AUGUSTUS and DEWAR, JAMES

73 Cotton gin
United States patent on 14 March 1794 (US patents were not numbered at this time)   Cotton gin. WHITNEY, ELY

74 Crease-resisting fabrics
304,900 (1929)   An improved cellulosic fabric and the production thereof. FOULDS, R. P. and MARSH, J. P., Tootal Broadhurst Lee Co. Ltd

75 C.S. gas
967,660 (1964)   Improvements in apparatus for controlling riots. FINN, DOUGLAS HART and others, War Department

76 Cylinder boring
1,063/1774   Casting and boring iron guns or cannon. WILKINSON, JOHN

An urgent need for an accurate cylinder boring machine arose with the invention of the Newcomen and Watt engines in the first half of the eighteenth century. The first sufficiently accurate machine was built by Wilkinson at Bersham Ironworks in or about 1775. This consisted of a long cylindrical bar mounted in bearings at each end, fitted with a cutter head which could be traversed along the bar from end to end. This machine was never patented. The gun boring machine described in the patent specification above is sometimes confused with it but it was different in principle. It may, however, have deterred others from copying the larger machine since there is no record of any such machine being erected other than by Wilkinson until after this patent had been revoked in 1779.

77 Dam-buster bomb

937,959 (1963) Improvements in explosive missiles and means for their discharge. WALLIS, BARNES NEVILLE, Vickers Aircraft Ltd

Means are provided to support an explosive missile and impart a spinning motion thereto about a horizontal axis at right angles to the direction of attack together with release gear to free the missile from the support and means to positively propel the missile towards the target. The patent was filed in 1942.

77 These drawings are copied from specification 937,959. The delay between the filing of the application and its acceptance and publication was seemingly for security reasons. No patent applications thought to be of interest to potential enemies of the State will be published.

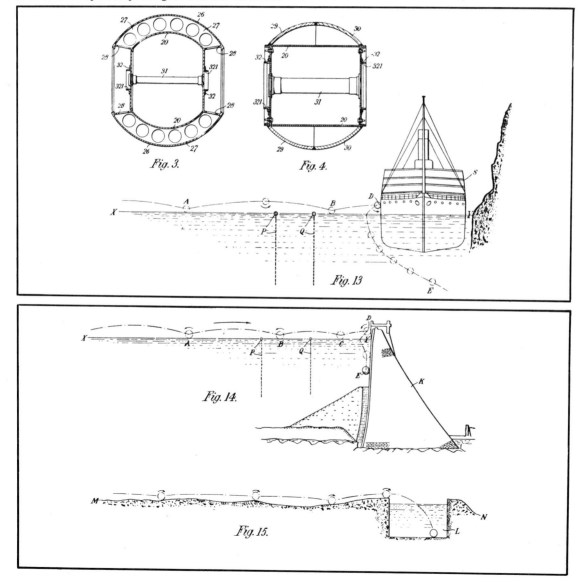

Fig. 3.

Fig. 4.

Fig. 13

Fig. 14.

Fig. 15.

78 DDT (dichlorodiphenyltrichloroethane)

547,871 (1942) Manufacture of aqueous emulsions for insecticide purposes. Hughes, Rosling Morgan for J R GEIGY A–G, SWITZERLAND

DDT was first prepared by OTHMAR ZEIDLER in 1874. Its properties as an insecticide were discovered by PAUL MÜLLER of Geigy in 1939. The discovery was notified to the Allies, via the British Legation, in 1942 and was used by them extensively in the later years of the war.

79 Deacon meter

937/1873 Apparatus for measuring the flow of water in pipes. DEACON, GEORGE FREDERICK. See also 4,264/1873.

80 Decca navigation system

584,727 (1947) Improvements in radio control systems for movable objects. SCHWARZ, HARVEY FISHER and O'BRIEN, WILLIAM JOSEPH (Decca Radio and Television Ltd)

O'Brien, an American engineer, was the first to consider the basic concept of this system in 1937. He failed at that time to interest the US Navy, the Civil Aeronautics Authority or American industry. In 1939 he wrote to Schwartz, another American, who was working for Decca in England. He was able to interest A. V. Alexander, then First Lord of the Admiralty, sufficiently for a demonstration to be arranged. This was immediately successful and the system was put into operation for use by minesweepers on the Normandy landings in June 1944. When the system came off the official secrets list after the war a committee under Sir Henry Tizard opined that the system was too limited in application to be commercially viable. The Decca group immediately set up a subsidiary company to run an experimental chain of stations in south-east England in 1946; these soon proved their worth during the extreme winter of 1946–47 in assisting colliers to bring in essential supplies of coal in weather so bad that seagoing vessels not equipped with Decca equipment were unable to get through.

81 Degaussing

532,795/1939 Protective means for ships against mines and torpedo attack. BARNES, FRANKLIN

82 Dental drill

1,017/1864 Apparatus for drilling, cutting and polishing teeth. HARRINGTON, GEORGE FELLOWS

> A.D. 1864, April 22.—N° 1017.
> HARRINGTON, GEORGE FELLOWS.—"Improvements in machinery or apparatus for drilling, cutting, grinding, and polishing teeth whilst in the mouth." These are "the substitution of any suitable arrangement of spring clockwork for the archimedean screw or bow or other contrivance hitherto employed for giving a rotatory motion to tools intended to operate upon teeth whilst in the mouth." The clockwork "is contained within a hollow metal box, case, or holder of a convenient size to be held in the hand of the operator." "When the spring is wound up the clockwork immediately commences to revolve, and thus imparts rapid rotatory motion to the tool." "The revolutions of the tools are arrested or controlled by the pressure of the thumb or finger, or of a break lever against a break wheel, which is fitted on to the arbour or spindle of one of the train of wheels forming the clockwork."
> [Printed, 10d. Drawing.]

82 This is copied from one of the early Woodcroft abridgment volumes for the Class relating to 'Medicine, Surgery and Dentistry'.

83 Dentures

1,803 (1791) Composition for making artificial teeth; springs for fastening the same. CHEMANT, NICHOLAS DUBOIS DE

84 Detergents

Sodium reduction process 14,758/1903 Manufacture of alcohols and alcohol derivatives applicable in the manufacture of perfumes, flavourings or the like, and of primary alcohols generally. BOUVEAULT, L. and BLANC, G.
Hydrogenation of fatty acids into fatty alcohols 1,515/1903 Process for converting unsaturated fatty acids or their glycerides into saturated compounds. NORMAN, W.; BOEHME, H. TH. and SCHRAUTH, WALTER, Deutsche Hydrierwerke

Detergents did not come into general use until after the Second World War. A large number of patents were taken out during the period from 1925 onwards. It is difficult to pinpoint the most significant of these. The above two patents relate to some early basic processes in their development. See also under ENZYME WASHING AGENTS.

85   Diesel engine

7,241/1892   A process for producing motive work from the combustion of fuel. DIESEL, RUDOLF. See also 4,243/1895 and 7,657/1898.

Many features of the Diesel engine, including compression ignition, were anticipated in the engine of HERBERT ACKROYD STUART, 7,146/1890, which was itself designed mainly with the aim of preventing the pre-ignition troubles of earlier engines. Diesel, however, must be given credit for his contribution to the fuller understanding of the thermodynamic principles involved. Later developments were related for the most part to improvements in fuel injection systems. The first practical air-less fuel injection was that due to JAMES McKECHNIE, 27,579/1910. This was used by the Royal Navy in the 1914–18 war and gradually replaced the original air-blast system. Pressure-charging of two-stroke engines was introduced by the Swiss ALFRED BUCHI, 24,980/1906. There was a rapid development of high-speed Diesel engines in the 1930s when they began to be adopted more widely for road vehicle propulsion. Again the key was improvements in fuel injection methods. The ROBERT BOSCH COMPANY of Stuttgart began a new trend by setting up as specialist manufacturers of fuel injection equipment and their designs (see for example United States 1,831,649 (1931), BAUER, OTTMAR) were sufficiently improved that they were able to sell them to various of the engine manufacturers who before had designed and built their own injection systems.

86   Disc brakes

26,407/1902   Improvements in the brake mechanism of power-propelled road vehicles. LANCHESTER, FREDERICK WILLIAM

This was the first patent for a disc brake of the 'spot' type. The first known patent for a clutch-type brake was 14,495 (1908), HOLZE, ERNEST. The latter was designed to brake horsedrawn carriages in the event of the horse bolting. The first commercially available disc brake for motor vehicles which preceded the presentday trend to disc brakes was described in 688,382 (1953), WRIGHT, JOSEPH and BUTLER, JAMES HENRY, Dunlop

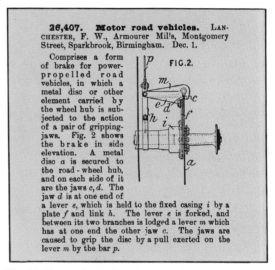

26,407. Motor road vehicles. LANCHESTER, F. W., Armourer Mills, Montgomery Street, Sparkbrook, Birmingham. Dec. 1.

Comprises a form of brake for power-propelled road vehicles, in which a metal disc or other element carried by the wheel hub is subjected to the action of a pair of gripping-jaws. Fig. 2 shows the brake in side elevation. A metal disc *a* is secured to the road-wheel hub, and on each side of it are the jaws *c, d*. The jaw *d* is at one end of a lever *e*, which is held to the fixed casing *i* by a plate *f* and link *h*. The lever *e* is forked, and between its two branches is lodged a lever *m* which has at one end the other jaw *c*. The jaws are caused to grip the disc by a pull exerted on the lever *m* by the bar *p*.

86   Abridgment for the Lanchester disc brake.

Rubber Company Ltd. For a self-servo disc brake of simple design, see 713,797 (1954), WINGFOOT CORPORATION.

87   Diving bell

279 (1691)   A certain new engine or instrument whereby by conveying of air into the diving vessel they can maintain several persons at the same time to live and work safely under water at any depth for many hours together. EVANCE, SIR STEPHEN; TYSSEN, FRANCIS; HOWLAND, JOHN and HALLEY, EDMUND

Halley was not the originator of the diving bell which was known even as early as the time of Aristotle. Francis Bacon wrote of diving vessels being used for working on submerged wrecks and there are reports of their use for lifting a wreck from the sea bed in 1642. Halley's contribution was to delineate carefully the defects of the simpler bells and to prescribe remedies; in particular, he provided means for supplying fresh air to the bell and releasing foul air. The fresh air was carried down in weighted barrels which had a hole in the base and a length of flexible tubing connected to a second hole in the top. The end of the tubing was bent over to below the base of the barrel while the barrel was lowered but was then turned up to release air in the bell as water entered the base hole. JOHN SMEATON, the well-known civil engineer, was the first to pump air to a diving bell.

88   Driving chains
2,515/1864   Improvements in toothed chains for working toothed or chain wheels. SLATER, JOHN. See also under BICYCLES.

89   Droop-snoot nose for aircraft
723,895 (1955)   Improvements relating to aircraft. FAIREY AVIATION CO. LTD

This configuration has since become well known through its application to the Concorde.

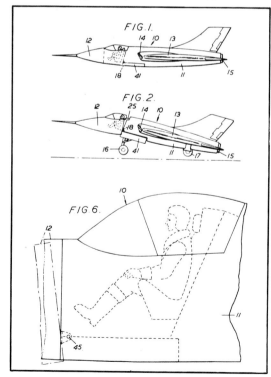

89   Droop-snoot nose.

90   Dual spectrum copying process
1,002,350 (1965)   Heat sensitive copy sheets and methods of using them. MINNESOTA MINING AND MANUFACTURING COMPANY

This process enables thermographic materials to produce long-lasting images by placing one heat-sensitive reactant in one sheet with a light sensitive dye which, on exposure to heat, prevents reaction of the reactant with a second heat-sensitive reactant in a separate sheet. Since materials are cheap and the apparatus simple, the process produces cheap copies, so many applications of the process are being developed.

91   Dye transfer
2,799/1871   Improvements in photomechanical printing, and in apparatus to be used in such printing, parts of which apparatus are also applicable to other purposes. EDWARDS, ERNEST

This suggests that if the water that is absorbed by the untanned portions of a photographic gelatin transfer plate making it ink repellent contained aniline dyes both ink and dye would transfer. This is probably the first disclosure of dye transfer.

92   Dynamite (Nobel's safety powder)
1,345/1867   Improvements in explosive compounds and in means of igniting same. Newton, William Edward for NOBEL, ALFRED

Nitroglycerine was discovered by Sombrero in 1846, but it was Nobel who made possible its safe manufacture. He erected the first nitroglycerine factory at Helenborg, near Stockholm in 1862. In 1864 he found means of firing nitroglycerine by detonation rendering it useful as a blasting explosive. However, in its liquid form it was much too dangerous to handle and only after much experimentation did Nobel discover that Kieselguhr, a siliceous earth, could be used to absorb the nitroglycerine so that it might be handled with a greater degree of safety while sacrificing only a small part of its effective strength.

93   Egg cartons
215,670 (1924)   Improvement in and relating to

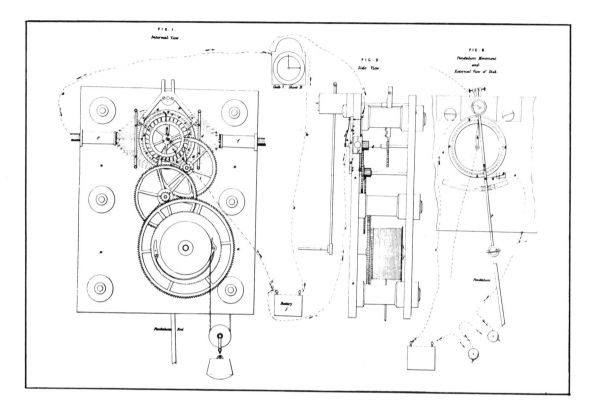

carton egg carriers. CANTELON, DAVID ARTHUR and
ZUEFLE, JOHN

One of the early examples of a method of forming
an egg carton from a flat blank.

94   Electric candle

3,552/1876   Improvements in electric light.
Applegarth, Robert for JABLOCHKOFF, PAUL

Two carbon rods were laid parallel, with insulating
substance between them. The rods extended
beyond the insulator and an electric spark across
them gave the required light. As the carbon rods
were consumed, so was the heated end of the
insulator. The first demonstration of these lamps
for street lighting was on the Avenue de l'Opera,
Paris, in 1878. See also 494 and 1,997 of 1877.

95   Electric clock

8,783 (1841)   Application of moving power to
clocks and timepieces. BAIN, ALEXANDER and
BARWISE, JOHN

95   Drawings of Bain's electric clock specification 8,783.
WHEATSTONE also claimed to have invented the
electric clock. There is no doubt that he had
worked on it and had advised Bain to stop work on
his clock when Bain approached him and his
partner for financial support in 1840. The first
clock to be driven by an alternating current is
described in 125,766 (1919), WARREN CLOCK
COMPANY. The first clock to apply electromagnetic
impulses to a balance wheel as opposed to a
pendulum was named the 'Eureka' 14,614/1906,
Haddam, Herbert John for KUTNOW BROTHERS,
New York. In this a straight solenoid with a soft
iron core was mounted on the balance wheel and
this oscillated between the poles of a horseshoe
magnet. The solenoid was energised on every
alternate swing by a simple contact system and an
impulse was given to the balance wheel. A similar
system was adopted in the first electric watch to be
placed on the market: the American 'Hamilton
500' of 1957. For later electronic watches see
TUNING FORK WATCHES.

96   Electric incandescent lamp
United States 223,898 (1880)   Electric lamp.
EDISON, THOMAS ALVA
4,933/1880   Improvements in electric lamps, and
the materials used in their construction. SWAN,
JOSEPH WILSON

The filament of the Edison lamp was a spiralled
thread of carbon having a resistance of several
hundred ohms as compared to the four ohms or so
of earlier carbon lamps. Platinum wires connected
the ends of the filament to lead-in conductors
sealed in a glass bulb. The bulb was exhausted and
hermetically sealed after a vacuum had been
established. It was the first bulb to be sufficiently
stable to be a commercial success in the United
States and was successfully defended in a number
of court cases. Swan also used hermetically sealed
bulbs of glass. His main contribution was in the
filament which was made by treating carbon-
impregnated cotton with sulphuric acid. In later
developments of such filaments Swan squirted
nitrocellulose solution through small holes
anticipating the development of artificial textile
fibres. See also p. 22.

97   Electric motor
United States 132 (1837)   Improvements in
propelling machinery by magnetism and
electromagnetism. DAVENPORT, THOMAS

This is the earliest patent for an electric motor.
Davenport produced several types of motor at
around this time and applied some of them to the
drilling of iron and steel and turning hardwood.

98   Electricity distribution
4,942 (1881)   An improved method of applying
electric currents in the production of light, and
lamp employed therefore. Pitt, Sydney for
GAULARD, LUCIEN and GIBBS, JOHN DIXON

This, the first successful alternating system, was
first demonstrated at the Electrical Exhibition in
the Westminster Aquarium in 1883. The following
year an experimental circuit was installed to give
arc and incandescent lighting in some passenger
stations of the Metropolitan Railway, London.
Transformers were used in this experimental
circuit. Another early patent of interest which
proposed the use of transformers in an alternating
distribution system is German patent 33,951
(1855), DERI, MAX

99   Electrolytic machining
335,003 (1930)   Method and apparatus for the
electrolytic treatment of metals. GUSSEFF,
WLADIMIR

This is the earliest known patent for making holes
or cavities in metals by localised electrolytic
erosion using a cathode tool. The technique has
become important with its application to the
machining of hard refractory metals and alloys.

100   Electron microscope
German patent DRP 485,155 (1929)   Method
and equipment for the automatic detection,
measurement and counting of individual particles
of any kind, shape and size. STINTZING, H.

402,781 (1933)   Improvements in or relating to
cathode ray tubes RÜDENBERG, R., Siemens
Schuckertwerke, Aktiengesellschaft

It was in 1883 that H. HERTZ found that cathode
rays could be deflected by a magnetic field. In
1897 J. J. THOMPSON refined Hertz's experiments
and postulated the existence of the electron.
Stintzing's patent contains the first published
proposals for the use of these discoveries in
microscopy but there were technical construction
difficulties that make it improbable that a working
instrument could be made to his design. The first
patent for an electron lens was taken out in
Germany, DRP 690,809 (1929). RÜDENBERG was
the first to file patents for comprehensive workable
electron microscopes. The British patent noted
above is for an electron lens only; other patents
were filed by Rüdenberg in Germany, France,
Austria, Switzerland and the United States. The
German patent DRP 895,635 was filed first but
there was considerable delay in its acceptance and
publication. It was published in 1953 with a
granting date of 1931. The French patent 737,716
(1932) was the first to be published.

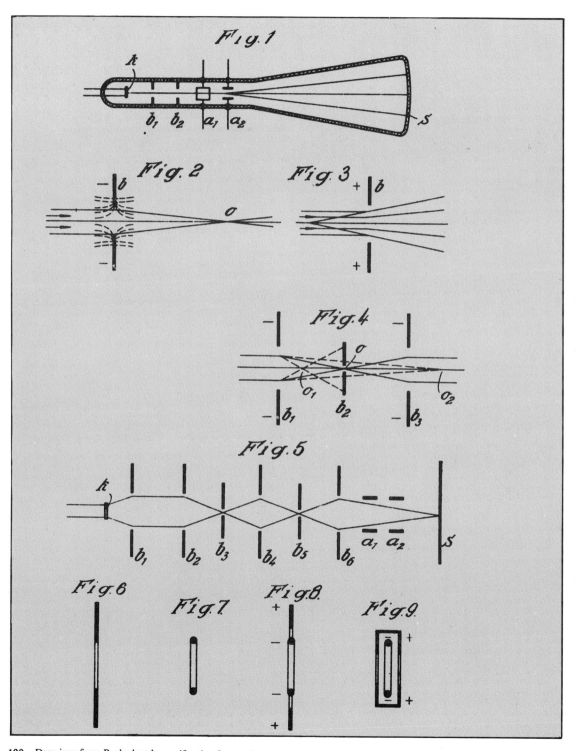

100 Drawings from Rudenberg's specification for an electron lens.

PATENT SPECIFICATION
NO DRAWINGS

1,156,238

Date of Application and filing Complete Specification: 25 April, 1967.
No. 18929/67.
Application made in United States of America (No. 544846) on 25 April, 1966.
Application made in United States of America (No. 620608) on 6 March, 1967.
Complete Specification Published: 25 June, 1969.
© Crown Copyright 1969.

Index at acceptance: —C5 D(6A2, 6A5A, 6A5B, 6A5C, 6A5D1, 6A5D2, 6A5E, 6A6, 6A8B, 6B1, 6B2, 6B3, 6B4, 6B5, 6B6, 6B7, 6B8, 6B10A, 6B10C, 6B11A, 6B12A, 6B12B1, 6B12B2, 6B12B3, 6B12C, 6B12E, 6B12F1, 6B12F2, 6B12F4, 6B12G1, 6B12G2A, 6B12G3, 6B12G4, 6B12K1, 6B12K3, 6B12L, 6B12M, 6B12N1, 6B12N2, 6B12N4, 6B12NX, 6B12P, 6B13, 6B15, 6C6, 6C8, 6D)

Int. Cl.: —C 11 d 7/42

COMPLETE SPECIFICATION
Conglutination of Enzymes and Detergent Compositions

We, THE PROCTER & GAMBLE COMPANY, a Company organised under the laws of the State of Ohio, United States of America, of 301 East Sixth Street, Cincinnati, Ohio, United States of America, do hereby declare the invention, for which we pray that a patent may be granted to us, and the method by which it is to be performed, to be particularly described in and by the following statement: —

This invention relates to enzyme-containing detergent compositions and to a process for conglutinating enzymes and detergent compositions.

Enzymes have been used as cleaning aids for many years. As early as 1915, Rohm found that fabrics could be cleaned more easily and at lower temperatures when pretreated with fat and protein digesting enzymes. See Rohm, Germany patent specification No. 283,923. Later, in, 1932, enzymes were utilized in a soap composition having greatly improved cleansing action. (See U.S. patent specification No. 1,882,279). Enzymes aid in laundering by attacking soil and stains found on soiled fabrics. Soils and stains are decomposed or altered in such an attack so as to render them more removable during laundering.

Enzymes can be used either in a soaking or pre-wash product designed to prepare soiled fabric for more effective detergency when the fabrics are subjected to conventional laundering, or else as a component of a detergent formulation containing conventional cleaning ingredients. The enzymes suitable for such laundry uses are usually found in a fine powder form. Enzymes are expensive and powerful materials which must be judiciously formulated and used. Such fine powders of concentrated materials are difficult to handle, difficult to measure and difficult to formulate.

Prior art enzyme-containing laundry products are mechanical mixtures of a fine enzyme powder and other granular materials. Enzyme powder in such mechanical mixtures tends to segregate, resulting in a non-uniform product. Non-uniformity results in an undependable product in use, particularly for measurement purposes. Such mechanical mixtures also present stability problems resulting from the mobility of the enzyme powder in the mixture; it is exposed to some cleaning ingredients and environmental conditions which may either attack the enzyme or aid it in degrading itself. For example, moisture tends to cause the enzyme to degrade itself; many enzymes are incompatible with highly alkaline detergent materials such as caustic soda, particularly in the presence of moisture.

Accordingly, an object of this invention is to provide a granular detergent composition that contains enzymes. Another, more specific, object of this invention is to provide an enzyme-containing detergent composition wherein the enzymes remain uniformly distributed throughout the product and the stability of the enzymes in the detergent composition is enhanced. Another object of this invention is to provide a process for incorporating enzymes into a detergent composition which minimizes the contact of the enzymes with materials which are deleterious to the enzyme.

Further objects will become apparent from the detailed description given hereinafter. All parts, percentages and ratios set forth

[Price 4s. 6d.]

106   This is the first page from one of the biological washing powder specifications. From the 'prior art' statement, lines 15–61, we see that the use of enzymes for washing purposes was contemplated as early as 1915. Note that some 'objects' of the invention are given from line 62 onwards.

101   Electronic clock
995,546 (1965)   Improvements in or relating to electronic clocks. P. VOGEL ET CIE

The first fully electronic clock with no moving parts and with cathode ray tube time display.

102   Electronic counter
436,420 (1935)   Improvements in or relating to electrical circuits employing gaseous discharge tubes. STANDARD TELEPHONES AND CABLES LTD.

The first electronic counter.

103   Electronic watch
1,057,453 (1967)   Electronic timepieces. P. VOGEL ET CIE

The first fully electronic integrated circuit watch, with electroluminescent display.

104   Eniac
709,407 (1954)   Electronic numerical integrator and computer. ECKERT-MAUCHLY COMPUTER CORPORATION

The first all-electronic computer.

105   Envelope making machine
10,565 (1845)   Manufacture of envelopes. HILL, EDWIN

Edwin Hill was the brother of SIR ROWLAND HILL who instigated the penny post and created a demand for envelopes. The invention of the envelope itself is often credited to S. K. BREWER, a Brighton bookseller.

106   Enzyme washing agents
First to note washing properties   German 283,923 (1915)   Verfahren zum Reingen von Waschestucken aller Art. ROHM, OTTO
Incorporation in a soap compound   United States 1,882,279 (1932)   Process of making a soap compound. FRELINGHUYSEN, GEORGE G.

Attachment of enzyme compounds to detergent granules   1,156,238 (1969)   Conglutination of enzymes and detergent compositions. PROCTER AND GAMBLE COMPANY

1,151,748 (1969)   Granular enzyme-containing laundry composition. PROCTER AND GAMBLE COMPANY

107   Epsom salts
354 (1698)   The way of making the salt of the purgeing waters perfectly fine in large quantities and very cheape, so as to be commonly prescribed and taken as a general medicine in this our kingdom. GREW, NEHEMIAH

This is the first recorded medicinal patent.

108   Evaporated milk
11,703 (1847)   Treating milk for the purposes of nutriment. GRIMWADE, THOMAS SHIPP
United States 15,553 (1856)   Concentration of milk. BORDEN, GAIL

Borden is usually given credit for the origination of the presentday evaporated milk industry. Grimwade, however, was marketing an evaporated milk product as early as 1848.

109   Evaporation and sugar refining
3,754 (1813)   Preparing and refining sugars. HOWARD, EDWARD CHARLES
United States 3,237 (1843)   Improvements in sugar works. RILLIEUX, NORBERT

Howard's patent was the first to propose evaporation under vacuum. Rillieux is credited with being the first successfully to operate the process of multiple effect evaporation in which the liquid is passed through a series of closed vessels each of which is maintained at a higher vacuum than the preceding one; see 5,296 (1880). This system which involves the multiple utilisation of latent heat in a most economic manner was first suggested in 5,394 (1826) CLELAND, WILLIAM, and developed by the French, see for instance French patent 12,388 (1843) DEGRAND, JEAN ALEXANDRE ELZEAR.

110   Facsimile
9,745 (1843)   Production and regulation of electric currents, electric time-pieces, and electric printing and signal telegraphs. BAIN, ALEXANDER
This patent described the basic concepts of

facsimile for the first time. In the form described it never came into general use. The first commercial facsimile service was set up in France in 1860 and used a system developed by an Italian, 2,395/1861, CASELLI, GIOVANNI. This incorporated the principles laid down by Bain and some of the improvements shown in 12,352 (1848), BAKEWELL, FREDERICK. An interesting development aimed at better synchronisation was the use of tuning forks as in 1,920/1869 by the Parisian LUDOVIC D'ARLINCOURT. The first to propose the use of electric motors in facsimile machines was WILLIAM SAWYER, United States 171,051 (1875). The transmission of half-tone pictures was first successfully demonstrated by NOAH S. AMSTUTZ of Cleveland in 1891.

111   Ferromanganese (Spiegeleisen) use in steel manufacture
8,021 (1839)   Manufacture of iron and steel.
HEATH, JOSIAH MARSHAL

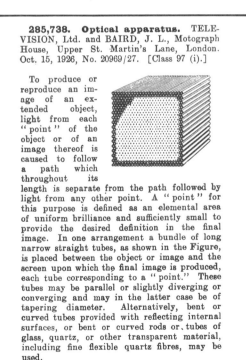

285,738. **Optical apparatus.** TELE-
VISION, Ltd. and BAIRD, J. L., Motograph
House, Upper St. Martin's Lane, London.
Oct. 15, 1926, No. 20969/27. [Class 97 (i).]

To produce or reproduce an image of an extended object, light from each " point " of the object or of an image thereof is caused to follow a path which throughout its length is separate from the path followed by light from any other point. A " point " for this purpose is defined as an elemental area of uniform brilliance and sufficiently small to provide the desired definition in the final image. In one arrangement a bundle of long narrow straight tubes, as shown in the Figure, is placed between the object or image and the screen upon which the final image is produced, each tube corresponding to a " point." These tubes may be parallel or slightly diverging or converging and may in the latter case be of tapering diameter. Alternatively, bent or curved tubes provided with reflecting internal surfaces, or bent or curved rods or, tubes of glass, quartz, or other transparent material, including fine flexible quartz fibres, may be used.

112   A bright idea for an application of fibre optics that never quite worked out in practical terms.

112   Fibre optics
285,738 (1928)   Production of optical images.
BAIRD, JAMES LOGIE, Television Ltd
United States 1,751,584 (1930)   Picture transmission. HANSELL, CLARENCE W., Radio Corporation of America

The first recorded scientific demonstration of fibre optics principles was given by JOHN TYNDALL at the Royal Society in 1870. It is possible that the basic principles were known by Greek and Venetian glass blowers well before then. The two patents above, both related to the early development of television, are the first references to practical applications. The present resurgence of interest in practical applications began in the early 1950s with attempts by A. C. S. VAN HEEL in the Netherlands and H. H. HOPKINS and N. S. KAPANY at Imperial College, London, to develop a 'flexible fibrescope'. The first use of the term 'fibre optics' was in an article by Kapany in 1956.

113   Film cartridges
375,912 (1932)   Improvements in or relating to film charging devices for use with cinematographic cameras and projectors. HILLERY-COLLINGS, CHARLES EDWARD

First disclosure of a coaxially arranged double reel film cartridge as used for 'Super 8' cartridges.

114   Film projection
1,110,538 (1968)   Film threading device. McKEE, EDWARD SHERMAN, Eastman Kodak Company

The first device using a driven belt and stripping finger for removing film from a reel or cartridge in order to feed it to a cinematograph projector.

115   Fire-proofing fabrics
551 (1735)   Preventing paper, linen, canvas, etc, from flaming or retaining fire. WYLD, OBADIAH

116   Fish fingers
257,222 (1927)   Improvements in methods of preparing fish foods. BIRDSEYE, CLARENCE

See also under RAPID FREEZING.

PATENT SPECIFICATION

Convention Date (United States): Aug. 24, 1925.

Application Date (in United Kingdom): March 8, 1926. No. 6442/26.

Complete Accepted: April 14, 1927.

257,222

COMPLETE SPECIFICATION.

Improvements in Methods in Preparing Fish Foods.

I, CLARENCE BIRDSEYE, of 35, Mount Pleasant Avenue, Gloucester, Massachusetts, United States of America, citizen of the United States of America,
5 do hereby declare the nature of this invention and in what manner the same is to be performed, to be particularly described and ascertained in and by the following statement:—
10 This invention relates to an improved method of preparing fish foods.

The objects of the invention are to provide an improved method of preparation of food products which will render the
15 same capable of being readily handled without damage, and permanent in form when sliced, cooked or otherwise treated after purchase and in preparation for eating; to enable scraps and other edible
20 portions of the food to be made saleable and attractive as well as convenient to handle, cook and use; to particularly adapt the invention to recovery and use of edible portions of otherwise wasted
25 fish; to avoid cooking or other expensive operations during manufacture; to obtain a homogeneous mass; to utilize inherent properties of the food in carrying out the process and thus avoid inclusion of
30 foreign matter; to provide a product which will resist disintegration even more positively than usual fish fillets while being cooked; to avoid the necessity of heavily salting the fish in preparing the
35 product; to avoid the necessity of tying, boxing and other artificial means for maintaining the shape of the food product; to secure simplicity and lower cost of construction and process of manufac-
40 ture; and to obtain other advantages and results as may be brought out in the following description.

According to this invention, the process of preparing fish consists in the combina-
45 tion of reducing the flesh only of fresh fish to fragments and placing the same

[*Price 1/-*]

in moulds and solidifying into a homogeneous cake, the gluey content of the fish being retained in the solidified mass.

As one specific embodiment of the 50 invention as applied to fish, I employ the flesh of properly cleaned, fresh fish. I wish to clearly distinguish at this time between "fresh" fish and "salt", "cured" or "pickled" fish; these 55 terms having a very definite meaning in the fish industry. "Salt" fish are not only so heavily salted that they require to be freshened before use, but they are at least partially dried, both of which 60 characteristics distinguish such fish from "fresh" fish which are preserved in their natural state as near as possible. In the case of "salt", "cured" or "pickled" fish the preservative effect of 65 salt is largely depended upon to prevent spoilage; while any product to be truly "fresh" and not "salted" relies upon the same preservative methods as are or can be applied to the proper handling of 70 unprepared or fresh fish. One such method of preservation of fresh fish is by subjecting the same to low temperatures, and this method is preferably the one adopted in connection with the fish pre- 75 pared according to my invention.

The fish used are preferably taken as fresh from the water as possible, for the usual reasons, and the edible flesh portions thereof utilized in whole or in part 80 in carrying out the invention. It may be noted that the invention enables use of edible scraps of fish, such as would be ordinarily wasted in cutting fillets from the bone, and accordingly the invention 85 may be employed to use only those scraps or to employ the entire edible flesh portion of the fish if so desired. Also, let it be stated here that the fish are kept as cold as conveniently possible throughout 90 the process, but I do not mean that the process has to be carried out in a

---

116   This Clarence Birdseye patent contains no section describing any 'prior art', but the 'objects' of the invention are set out in lines 12 to 42. The consistory clause begins, typically with the phrase 'according to this invention' on line 43 and a description of a particular embodiment is given from line 50 onwards.

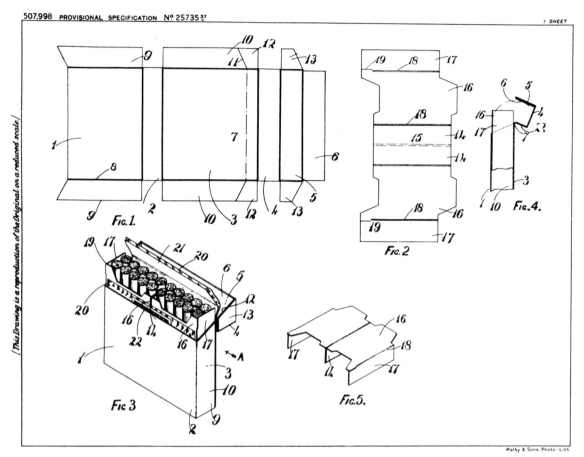

507.998 PROVISIONAL SPECIFICATION Nº 257353?

This Drawing is a reproduction of the Original on a reduced scale.

119 The drawings from provisional specification 507,998.

117   Flash bulbs
324,578 (1930)   Improvements in flash lights such as those used for photographic purposes.
OSTERMEIER, JOHANNES

This is the forerunner of the modern photographic safety flash bulb. It used aluminium wire or foil in oxygen. Earlier versions often exploded.

118   Flash cubes
1,034,090 (1966)   Photographic flash device.
PARSON, GEORGE WILLIAM and PETTS, RONALD GEORGE, Eastman Kodak Co.
1,204,973 (1970)   Percussively ignitable photoflash lamps. KOPELMAN, BERNARD, Sylvania Electric Products Inc.

The first of these patents was the first to disclose a flash cube in a cubical structure containing four

electrically ignitable flashlamps each having an associated reflector. The second patent is for the 'Magicube', a flash cube containing percussively ignitable flashlamps.

119   Flip-top cigarette packets
507,998 (1939)   Improvements in or relating to boxes. CHALMERS, JOHN WALKER, Mollins Machine Co. See also the companion patent 508,312 (1939).

These patents show an early example of the 'flip-top' cigarette packet which has now replaced the drawer and shell packet.

120   Float glass
769,692 (1957)   Improvements in or relating to the manufacture of flat glass. PILKINGTON, LIONEL; BETHUNE, ALEXANDER and BICKERSTAFF, KENNETH.

See also 19,829/1902 and United States patent 2,911,759 (1959).

In the float glass process liquid tin is used as the platform on which to form a ribbon of molten glass and to support the tin until the fire-finished surfaces set sufficiently to be handled without damage. Developing the process to success required a great deal of courage on the part of the Pilkington Company who spent eleven million pounds over seven years, almost entirely on the production of unsaleable glass before the process could be made commercially viable. One piece of unforeseen luck worked in their favour. The equilibrium between gravity and surface tension for the tin/glass system gave a ribbon which was 7 mm thick. Sixty per cent of the flat glass trade was in glass 6 mm thick which could easily be obtained from the 7 mm thick glass by traction. The process has since been found capable of further development, for instance, an electrofloat process has been devised which enables metallic ions to be electrolysed into the glass using the tin bath as a cathode. These are reduced to metal by the hydrogen present leaving a layer of metallic particles below the glass surface which improve the heat and light attenuating properties of the glass.

121   Float valves
730 (1758)   Fire engine for drawing water out of mines, draining lands and other purposes.
BRINDLEY, JAMES

Float valves are used today for a variety of purposes, the best known of which are the ball cock of the domestic water cistern and the control of petrol inflow to the automobile carburettor. The first known applications were associated with water clocks in the third century B.C. The first known reference to their use for domestic cisterns was in a building manual by WILLIAM SALMON, *The Country Builder's Estimator*, published in 1746. Brindley's patent above is for a steam engine that incorporated a float valve to regulate the water level in the steam boiler. This soon became a standard method of controlling water flow in steam engines. For another early example see 1,455 (1784), WOOD, SUTTON THOMAS.

122   Flour milling
2,054/1862   Roller mills for producing flour.
BUCHHOLZ, J. A. A.

Roller mills for grinding corn into flour date back as far as 1651. The first patent for roller milling is 675 (1753), WILKINSON, ISAAC. However, it was to be a long time before the roller mills became sufficiently successful to replace the older stone mills. In the early days the rollers were made of steel and the stones were not dispensed with altogether. The Buchholz system used both rollers and stones and was the first to be put into operation successfully in England, at the Albert Mills of Radford and Sons in Liverpool. Radfords claimed later to have installed a modified version of the Buchholz mill which was the first in England to dispense altogether with rollers. Roller milling did not come into its own until the introduction of porcelain rollers by F. WEGMANN of Naples. At first the process was kept secret and it was not until 1877 that porcelain rollers were introduced in England by HENRY SIMON in the A. McDougall's mill in Manchester, 1180 (1878).

123   Fluid logic device
French 788,140 (1934)   Procédé et dispositif pour dévier une veine fluide pénétrant d'autres fluides. COANDA, HENRI. The United States patent was 2,052,869 (1936).

These patents were the first to describe what is now known as the *Coanda effect* which forms the basis of a considerable amount of research which has been devoted to fluid logic devices since the late 1950s.

124   Fluorescent lamps
3,444/1903   Improvements in luminous gas or vapous electric lamps. Le Tall, Frederick William for COOPER-HEWITT ELECTRIC COMPANY

298,906 (1929)   Improvements in or relating to gaseous glow lamps. HULL, ALBERT WALLACE, The British Thomson-Houston Company Ltd
390,384 (1933)   Improvements in and relating to lamps. INMAN, GEORGE ELMER, The British Thomson-Houston Company Ltd
445,789 (1936)   Improvements in or relating to

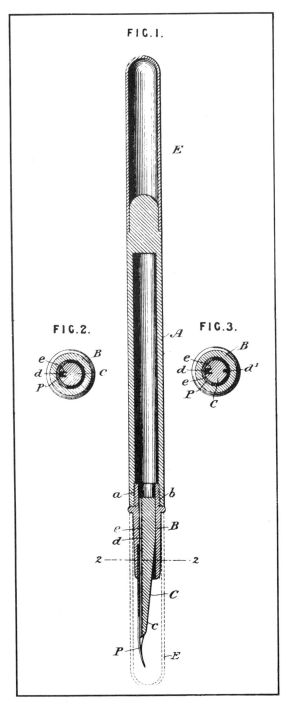

FIG. I.

E

FIG.2.   **FIG.3.**

A

e   *B*   *e*   *B*
d   *C*   *d*   *d'*
P   *e*
           P   *C*

a   *b*

e   *B*
d

2 ---   --- 2

C

c

P   *E*

126   The improvement claimed in this specification related to the configuration of ducts allowing air to flow into the reservoir space so that the flow of ink to the nib would be more reliable.

electric discharge devices provided with thermostatic switches of bimetallic strip type. RUFF, HAROLD ROBERT, The British Thomson Houston Company Ltd

The history of the development is one in which it is difficult to pick out a few significant steps. Gradual progress was made over many years with a series of small improvements with contributions coming from engineers and inventors from several countries. The Cooper-Hewitt lamp of the first patent above was the first low-pressure mercury discharge lamp, Hull's lamp, the subject of the second patent, marked one of the more notable developments in hot cathode designs pioneered by D. McFARLANE-MOOR and others. Inman's lamp was the first practical lamp using hot cathodes and fluorescent materials responsive to ultraviolet radiation and led to the lamps being marketed in the United States. The last of the patents above is one of a series relating to starting devices which preceded the first marketing of fluorescent lamps in England in 1939.

125   Flying shuttle
542 (1733)   Machine for opening and dressing wool; shuttle for weaving broadcloths, broad-baize, sail-cloths, or any other cloths, woollen or linen. KAY, JOHN

126   Fountain pen
3,125/1884   Improvements in fountain pens. Thompson, William Phillips for WATERMAN, LEWIS EDSON

This was probably the first workable pen. For earlier fountain pen patents see 3,214, 3,235 and 3,260 of 1809.

127   Four-stroke engine
2,081/1876   Improvements in gas motor engines. Abel, Charles Denton for OTTO, NICOLAUS AUGUST

128   Freeze-branding
1,107,149 (1968)   Method of cryogenic branding of animals. RESEARCH CORPORATION

This was the first patent to describe the branding of animals by the use of intense cold rather than by heat.

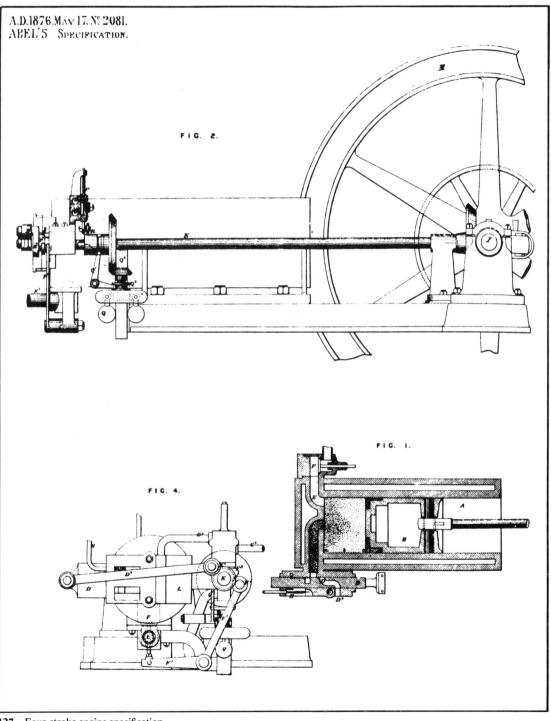

A.D.1876.May 17. N°. 2081.
ABEL'S Specification.

FIG. 2.

FIG. I.

FIG. 4.

127   Four-stroke engine specification.

129   Freon refrigerants
357,263 (1931)   Refrigeration. MIDGELEY,
THOMAS, jnr., HENNE, ALBERT LEON and McNARY,
ROBERT REED, Frigidaire Corporation

This is the first patent relating to the series of non-
toxic, non-inflammable organic compounds
containing fluorine.

130   Fuel cell
667,298 (1952)   Improvements relating to
galvanic cells and batteries. BACON, FRANCIS
THOMAS, British Electrical and Allied Industries
Research Association

131   Galvanised iron
7,355 (1837)   Improvements in coating or
covering iron and copper for prevention of
oxidation. CRAUFORD, HENRY WILLIAM

132   Gas containers
766,128 (1957)   Apparatus and method of
storing, shipping and using volatile hydrocarbons.
MORRISON, WILLARD LANGDON

This is the pioneer patent on methods of shipping
natural gas in large quantities and in liquid form.

133   Gas-discharge display panel
1,161,832 and 1,161,833 (1969)   Gaseous
display and memory apparatus. UNIVERSITY OF
ILLINOIS FOUNDATION

134   Gas mantle
15,286/1885   Manufacture of an illuminant
appliance for gas and other burners. VON
WELSBACH, CARL AUER

This was the first of Von Welsbach's patents for a
gas mantle. In practice it was not a great success.
It consisted of a cotton fabric base which had been
dipped in a liquid containing oxides of
zirconium, lanthanum and yttrium. A second
patent 3,592/1886, covered the use of oxide of
thorium either alone or mixed with other rare
metals. This was no more successful and it was by
chance that, while trying to obtain a pure thorium

nitrate, he discovered that a mixture of thorium
nitrate and cerium nitrate gave the best light. This
mixture was covered in England by 124/1893
taken out by JULIUS MOELLER, an employee of the
Welsbach Company.

135   Gas production
2,764 (1804)   Oven, stove, or apparatus for
extracting inflammable air, oil, pitch, tar and acids,
from all kinds of fuel and reducing the same into
coke and charcoal. WINSOR, FREDERICK ALBERT

JOHN BAPTIST VON HELMONT of Brussels was the
first to report experiments which yielded inflammable gas from charcoal. This was in 1609 and
it was many years later, in 1792, that WILLIAM
MURDOCK showed how gas could be produced in
sufficient quantities to be put to practical use. He
distilled the coal in an iron retort and piped the gas
to his house to provide light. He later installed gas
lighting at the Boulton and Watts works at Soho
in Birmingham. Murdock was one of several
investigators active at that time. Among the others
PHILIPPE LEBON obtained a French patent for
distilling coal in 1801. Winsor's main contribution
was on the commercial side. He was the first to
form a gas company and to propose the supply of
gas to many users from a central source.

136   Gears for bicycles
1,570/1896   Improvements in velocipedes and
auto-motor carriages. HODGKINSON, EDMUND HUGH
16,221/1901   Improvements in or relating to
variable speed gears for bicycles and other
machinery. STURMEY, JOHN JAMES HENRY
25,799/1905   Improvements in variable gear and
brake mechanism for velocipedes. ARCHER, JAMES,
The Three Speed Gear Syndicate Ltd

The Hodgkinson gear, marketed as the 'Gradient',
together with the four-speed gear 'Protean',
11,821/1894, LINLEY, A. V. were forerunners of the
derailleur gears which were later re-introduced to
England from France. The second and third of the
patents above are two of a series published from
1901 to 1906 leading to the well known Sturmey-
Archer gears.

137 Gelignite
4,179/1875 Explosive compounds. Newton, Henry Edward for NOBEL, ALFRED

138 Generator, electric
4,905/1876 Improvements in apparatus for producing the electric light, parts of which invention are applicable to other purposes. VARLEY, SAMUEL ALFRED

Several types of direct current generator were developed from 1830 onwards. This patent is generally recognised as the first to propose the use of compound field windings. The earliest alternating current generator was probably that constructed in Berlin by Siemens and Halske co-founders of the SIEMENS AND HALSKE COMPANY.

139 Geodesic dome (Fuller's)
963,259 (1964) Framework structure for buildings and the like. FULLER, RICHARD BUCKMINSTER

Geodetic structures for aircraft were invented well before this time. See, for instance, 429,188 (1935), WALLIS, BARNES NEVILLE, Vickers (Aviation) Ltd

140 Golf ball
17,554/1898 Improvements in balls for use in golf and other games. Boult, Alfred Julius for WORK, BERTRAM GEORGE

Describes a ball made mainly of windings of rubber thread around a central core. 4,165/1900, THE HASKELL GOLF BALL COMPANY, describes a machine for winding on the thread.

141 Gun cotton
11,407 (1846) Improvements in the manufacture of explosive compounds. Taylor, John for SCHONBEN, C. F.

142 Gunn effect oscillator
1,070,261 (1967) A semi-conductor device. GUNN, JOHN BATTISCOMBE, International Business Machines

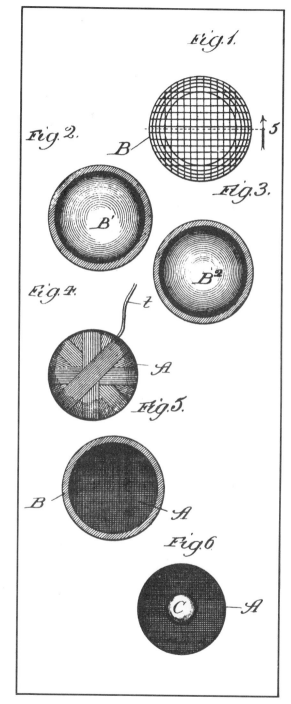

140 Until 1848 golf was played with leather balls stuffed with feathers. These were then superseded by balls made of solid gutta-percha. The Haskell balls replaced these progressively from 1902 onwards.

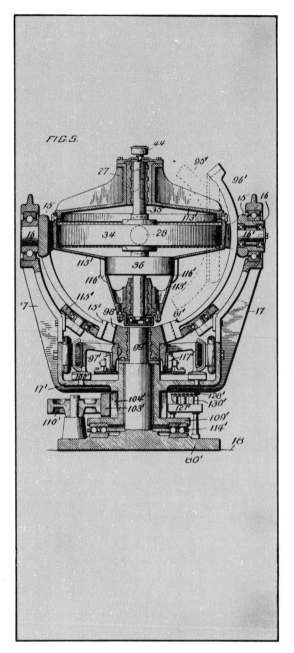

143 This is one of the drawings from Sperry's specification 11,513 (1908). Elmer Sperry's first attempts to commercialise the gyroscope were in 1907 when he attempted to interest circus proprietors in using a gyroscopic device as an aid for high wire performers. He then turned unsuccessfully to interest the automobile industry before turning to ship applications.

143   Gyroscopes
Gyrocompass 6,359/1905   Improvements in or relating to gyroscopes. ANSCHUTZ-KAEMPFT, HERMANN
Gyrostabilisers 11,513/1908   Gyroscopic apparatus for moving or controlling motion of bodies. SPERRY, ELMER AMBROSE. For ship stabilisers see 582/1914 or United States 1,150,311 (1915)
Marine gyropilot   United States 1,360,694 (1920)   Navigational apparatus. SPERRY, ELMER AMBROSE. See also 135,871 (1919).

144   Half-tone process
2,954/1865   Application of photography to the obtaining of printed proofs or impressions or engravings. BULLOCK, EDWARD and BULLOCK, JAMES

This patent was the first to propose a practical method of using a ruled screen for producing negatives for making photolithographic transfers. Some years before, in 1852, FOX TALBOT had set the scene by proposing the use of a crêpe or gauze screen between the positive and the bichromated gelatine during the exposure of photogravure plates to produce a grain or cellular structure. J. W. SWAN made further improvements to the process, 2,968–2,969 (1879). These early patents of Bullock and Swan described single-line screens. The first proposal for cross-lined screens came from the American F. E. IVES who later crossed the Atlantic to join Swan in setting up the Swan Engraving Company. The first newspaper printed using the half-tone process is reputed to be the New York *Daily Graphic* of 4 March 1880, the printer being STEPHEN H. HORGAN.

145   Hansom cab
7,266 (1836)   Cabs. GILLETT, WILLIAM STEDMAN and CHAPMAN, JOHN

The Hansom cab derived its name from an architect JOHN ALOYSIUS HANSOM who patented a safety cab in 1834 (6,733) and is said to have driven the prototype from Hinkley in Leicestershire to London. The cab design that later became so well known on the London streets contained many changes from Hansom's original

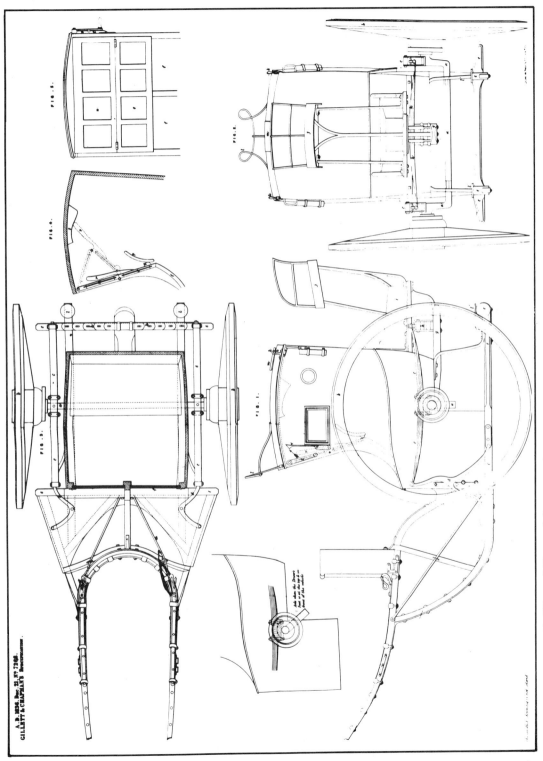

145 The Hansom Cab with Gillett and Chapman's improvements.

design. These changes, which involved the lightening of the main framework which carried the wheel axles and the provision of a better entrance platform making it practicable to drive the cab from the rear, were mainly the work of Chapman. During the time of the development Chapman was employed by a company called The Patent Safety Cabriolet Company, which had been formed specifically to work the Hansom patent, the rights in which they had bought.

146   Haymaking machine
680,537 (1952)   Device for laterally raking together crops or other vegetable products lying on the ground. VAN DER LELY, CORNELIS
Concerns a rake wheel type (finger wheel) haymaking machine. It is significant in that it was the subject of the longest case in patent litigation.

147   Helicopters
2,333/1859   Machinery or apparatus for navigating the air. BRIGHT, HENRY
327,381 (1930)   Improvements in or relating to aircraft with auto-rotative wings. CIERVA, JUAN DE LA, The Cierva Autogiro Company Ltd
551,156 (1943)   Improvements in or relating to aircraft. SIKORSKY, IVOR IVANOVICH, United Aircraft Corporation. See also 554,497 (1943).
614,676 (1948)   Improvements in or relating to rotary wing aircraft. Forsythe, Archibald Graham for DOBLHOFF, FRIEDRICH LIST

Bright's patent is the first recorded for a rotary wing aircraft although many expositions of the possibilities had been made before. The Frenchman L. C. BREGUET is credited with having been the first to produce a practical machine for manned flight in 1907. He applied for British patents but none of them appear to have been pursued to the stage of acceptance. Cierva's patent above relates to improvements to the autogiro first invented by him in 1923. Doblhoff was the first to produce a practical machine using jet-propelled rotors. Most of the development work was done in Austria during the war years. The Sikorsky patents are the forerunners of others that led to a long sequence of successful production machines in the United States. Sikorsky had been active in experimenting with rotary winged aircraft as early as 1909 when he was still in Russia but after a short time had moved on to the design of other types of aircraft and did not return to rotary wing studies until 1938.

148   Heliograph
3,390/1874   Improvements in apparatus for telegraphing by means of reflected light. MANCE, HENRY CHRISTOPHER

149   Hollerith card
327/1889   Improvements in the methods of and apparatus for compiling statistics. HOLLERITH, HERMAN

Hollerith was a twenty-year-old employee of the United States Census Bureau when he devised the first of his electric tabulating machines. About 100 such machines were used to tabulate the 1890 United States census rolls. The original machine was battery driven and included a sorting box which contained twenty-six compartments. The operator would insert a $5\frac{5}{8} \times 3\frac{1}{4}$ inch punched card into a press and lower a handle, spring actuated pins in the upper plate then made contact with pools of mercury in the lower plate wherever a hole was punched in the card. A completed circuit would open up a lid to one of the box compartments and cause one of the counting dials to advance one unit. The operator then manually placed the card into the compartment. By the time the 1900 census was due Hollerith had devised an improved machine with an automatic card feed very much like the modern card sorters.

150   Holograph
685,286 (1952)   Improvements in and relating to microscopy. GABOR, DENNIS, British Thomson-Houston Co., Ltd
1,104,041 (1968)   Wavefront reconstruction using a coherent reference beam. JURIS, UPATNIEKS and LEITH, EMMETT N., Battelle Development Corporation
1,142,702 (1969)   Interferometric examination of objects and materials. BURCH, MORRIS JAMES and others, National Research Development Corporation

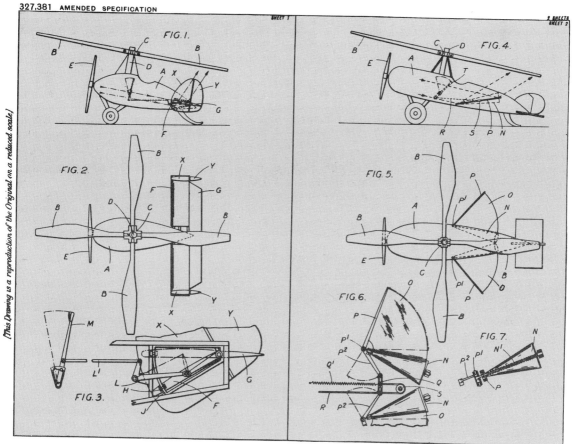

147 *Above* The drawings from Cierva's autogiro specification.

149 *Below* Hollerith could not possibly have foreseen how important his basic conceptions would become in peripheral and input equipment of the present-day electronic computer industry. These figures are from specification 327 (1889).

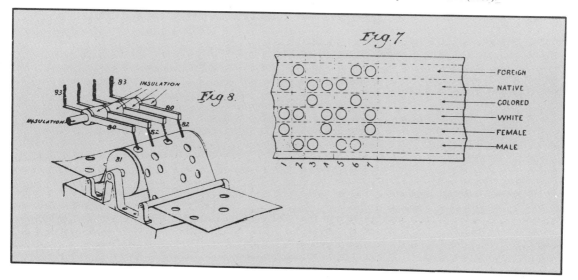

The original theory of holography was developed by Gabor. The method, however, remained largely theoretical until the advent of the laser as a coherent light source. The second of the patents above describes a method of off-axis holography as opposed to the on-axis method of Gabor. The off-axis method has been used in most applications of holography invented since. The last of the patents illustrates one of the many applications of holography, this one in the field of non-destructive testing.

151   Horseshoes, plastic
1,239,202 (1971)   Improvements in or relating to horseshoes. MAGUIRE-COOPER, RICHARD TERENCE

152   Hot-blast furnace
5,701 (1828)   Application of air to produce heat in fires, forges, or furnaces where bellows and other blowing apparatus are required. NEILSON, JAMES BEAUMONT

153   Hovercraft
854,211 (1960)   Improvements in or relating to vehicles for travelling over land and/or water. COCKERELL, CHRISTOPHER, Hovercraft Development Ltd. See also AIR CUSHION VEHICLES.

154   Hovis bread
12,736/1886   Improved treatment of wheatgerm and broken wheat. SMITH, RICHARD

The trademark *Hovis* derives from the two latin words *hominis*, meaning 'of man' and *vis* meaning 'strength'. Combined they sum up the idea that bread is the 'staff of life'. The trademark was chosen from hundreds of suggestions sent in to millers by their customers in a competition held after this new method of milling to preserve the wheat germ had been discovered.

155   H₂S radar display
599,889 (1945)   Improvements in radar navigating apparatus. THOMPSON, FREDERICK CHARLES, Ministry of Supply

This radar gives a pictorial display of the local terrain as seen from a moving craft and provides means for maintaining the display on the screen for a sufficient period of time for details to be distinguished, rather than allowing the display pattern to drift continuously as the craft moves.

156   Hub-dynamo for bicycles
468,065 (1937)   Improvements in electric generators. RAWLINGS, GEORGE WILLIAM, The Raleigh Cycle Company

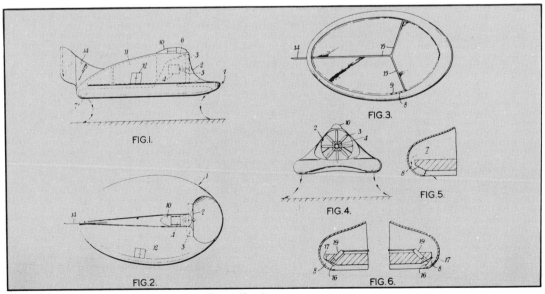

153   One sheet of the drawings from Cockerell's specification.

AMENDED SPECIFICATION

Reprinted as amended in accordance with the Decision of the Court of Appeal dated the nineteenth day of May, 1949, under Section 22 of the Patents and Designs Acts, 1946.

PATENT SPECIFICATION 468,065

Application Date: Nov. 26, 1935.   No. 32762/35.

Complete Specification Left: Nov. 26, 1936.

Complete Specification Accepted: June 28, 1937.

PROVISIONAL SPECIFICATION

Improvements in Electric Generators

We, THE RALEIGH CYCLE COMPANY LIMITED, of No. 177, Lenton Boulevard, Nottingham, a Company incorporated under the Laws of Great Britain, and GEORGE WILLIAM RAWLINGS, of " Willow Meer ", Park Hill, Kenilworth, Warwickshire, a British Subject, do hereby declare the nature of this invention to be as follows :—

This invention has reference to improvements in electric generators and is concerned more particularly with electric generators of the type wherein the magnetic field is created by a permanent magnet and which are required for generating current for lighting purposes on cycles.

With electric generators for use on cycles heretofore in use it has been necessary in order to obtain the desired output and steadiness of light for the rotor of the generator to be rotated at a considerably greater speed than the wheels of the cycle.

The present invention has for its object an improved electric generator of the aforesaid kind and primarily for the purpose specified which has a high output, which gives a steady light even at slow speeds but which is only driven at the same speed as the cycle wheels and which can be made sufficiently compact to be incorporated in the hub of a wheel.

The invention consists of an improved electric generator of the kind hereinbefore referred to comprising in combination a multipolar annular permanent magnet mounted within the wheel hub and rotatable therewith and a wound multi-pole-piece stator fixed to the hub spindle and disposed within the aperture of the annular magnet thus providing a generator in which all the rotating parts are fixed relatively to the rotating parts of the hub and all the

[Price 2/-]

stationary parts are fixed relatively to the hub spindle.

The invention further consists in an improved electric generator of the kind hereinbefore referred to wherein the magnet in the form of an annulus is made of a material having a high coercivity, said annulus being mounted within and rotatable with the hub of the wheel and wherein the stator is mounted upon the hub spindle and fixed relatively thereto and is wound to provide a relatively large number of pole-pieces preferably equal in number to the number of poles induced in the annular magnet.

The invention still further resides in the details of the improved electric generator to be described hereinafter.

A convenient embodiment of the invention will now be described in its application to an electric generator suitable for use with pedal cycles.

According to the said embodiment of the invention the hub of the cycle is made with an enlarged portion at one end and disposed within this enlarged portion and fixed thereto is an annular magnet of a material having a high coercivity such as one of the recently developed Nickel Alminium Alloys or one of the Cobalt magnet steels. This annular magnet which may conveniently have an external diameter of 1¼ inches and a width of ¾ inches is permanently magnetised so that a relatively large number of poles, say 20, is induced in the magnet. Mounted on the wheel spindle is the stator which is wave or lap wound so as to provide 20 radially disposed pole pieces which of course are located adjacent to the pole pieces of the annular permanent magnet said stator conveniently having a diameter of 3½ inches. The ends

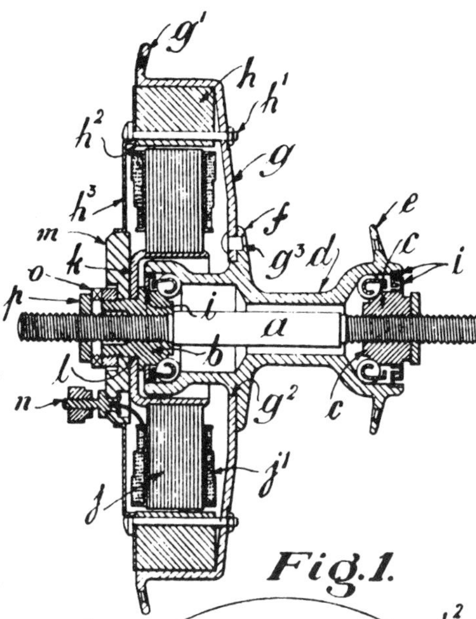

Fig.1.

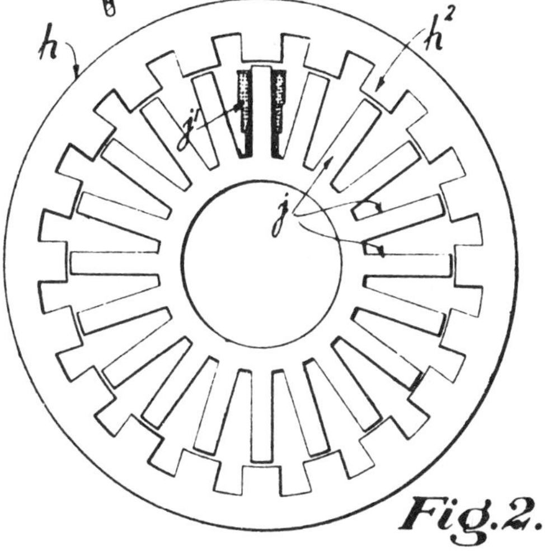

Fig.2.

156   There were some interesting court hearings over this patent in which there was much discussion over the precise meaning of various words and phrases used in the specification. What was a 'hub' or a 'steady light'? One judge went to great lengths in defining, for his own conscience, since neither of the protagonists had raised the issue, the exact meaning of the word 'cycle', i.e. did it include motor cycles, tricycles and pedal carts? The final outcome of the court actions hinged on the ambiguity of the word 'multipolar' (how many poles were permissible?) and four of the five claims were ruled to be invalid on this count. The fifth claim was allowed since it referred to a dynamo as described in the text and drawings, the drawings showed a 20 pole magnet.
It is fear of challenges for ambiguity that leads specification writers to be ultra-cautious.

157   Hydraulic press

2,045 (1795)   Certain new methods of producing and applying a more considerable degree of power in all kinds of mechanical apparatus and other machinery requiring motion and force, than by any means at present practised for that purpose. BRAMAH, JOSEPH. See also 3,611 (1812).

The principles for the application of hydraulic power had been enunciated well before this time. Bramah's most significant contribution was his use of cup leather packing to provide a seal between the walls of the cylinder and the moving piston.

158   Hydrodynamic couplings

13,864/1906   Improvements in and relating to the transmission of power. FÖTTINGER, HERMANN

This is the original specification disclosing hydrodynamic couplings and torque converters which form the basis for much of today's turbo-technology.

159   Ignition

5,784/1883   Gas or oil motors. Groth, Lorentz Albert for DAIMLER, GOTTLIEB

160   Induction furnace

Original furnace   700/1887 Improvements in electric furnaces and apparatus for heating, lighting, and carrying on chemical processes, and in the working of such furnaces or apparatus. FERRANTI, S. Z. DE
High-frequency induction melting   11,844 (1916)   Improvements in and relating to method of and apparatus for heating by electric induction. NORTHRUP, EDWIN FINCH, Ajax Metal Co.

161   Induction motor

6,481/1888   Improvements relating to the electrical transmission of power and apparatus therefore. Lake, Henry Harris for TESLA, NIKOLA

At the time of its introduction this was the only electric motor that would operate on alternating current. Engineers in other countries, notably GALILEO FERRARI in Italy, were working on similar motors and there were disputes over

priorities. For the synchronous induction motor see United States 694,092 (1902) DANIELSON, ERNEST.

162   Insulin

203,778/1922   A method of preparing extracts of pancreas, suitable for administration to the human subject. BANTING, FREDERICK and others

See also PROTAMIN INSULIN.

163   Intaglio printing

28,392 (1904)   Improvements in the methods of printing from photo-engraved intaglio plates. SWAN, JOSEPH WILSON and CAMERON-SWAN, DONALD

This sets out the basic principle of the deep-etched plates used for intaglio lithographic printing. United States 1,155,352 (1915) and Reissue 14,802 (1920) also cover plates having an intaglio image.

164   Internal combustion engine

4,315/1885   Improvements in motor engines worked by combustible gases or petroleum vapour or spray. DAIMLER, GOTTLIEB. See also 10,786 and 13,163/1885.

165   Invar

11,695/1897   Improvements in the manufacture of alloys of, or containing iron and nickel. Mewburn, John Clayton for LA SOCIÉTÉ ANONYME DE COMMENTRY-FOURCHAMBAULT

A non-expandable alloy of iron and nickel containing approximately 37 per cent of nickel.

166   Ion exchange

553,233 (1943)   Improvements in the treatment of solutions by ion-exchange methods. PEMBERTON, THOMAS ROLAND and others, The Permutit Co. Ltd

167   Jet propulsion engine

347,206 (1930)   Improvements relating to the propulsion of aircraft and other vehicles. WHITTLE, FRANK

This was Whittle's first patent relating to gas

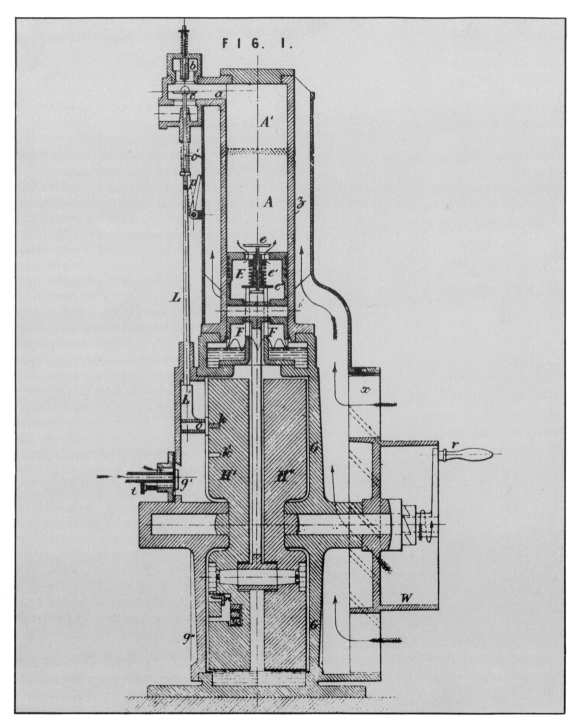

FIG. 1.

164 The improvement claimed in this specification involves the pumping of fuel mixture through the piston head into the chamber at the top and the bottom of the outstroke following the working instroke. The first assists in clearing the products of combustion while the second augments the amount of fuel in the chamber after the main fuel intake.

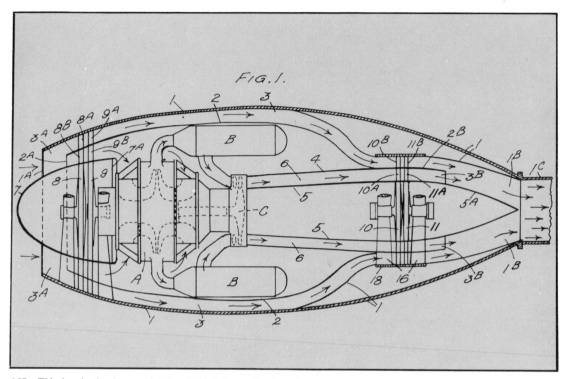

FIG.1.

167 This drawing has been copied from Whittles's specification 583,112.

turbine jet propulsion engines. The concept of the
bypass or ducted fan type jet propulsion engine
was first set out in 471,368 (1937) and developed
in 583,112 (1946). Jet propulsion for aircraft as
such had been proposed in principle long before,
see for instance 2,115/1867, BUTLER, JAMES
WILLIAM and EDWARDS, EDMUND.

168   Kaleidoscope
4,136 (1817)   New optical instrument called the
Kaleidoscope for exhibiting and treating beautiful
forms and patterns of great use in all the
ornamental arts. BREWSTER, DAVID

169   Lace loom
5,741 (1828)   Machinery for making bobbin-net
lace. LEVER, JOHN. See also 5,940 (1830) and 6,778
(1835).

Earlier machines were made by JOHN HEATHCOAT
who took out a number of patents between 1808
and 1843. His first successful machine is covered
by 3,216 (1809).

170   Lanoline
4,992/1882   Manufacture of fatty matter from
wool fat. GLASER, FRIEDRICH CARL

171   Latch needle
12,474 (1849)   Improvements in machinery for
the manufacture of looped fabrics. TOWNSEND,
MATTHEW and MOULDEN, DAVID

172   Lawn mowers
Early patent   2,859 (1805)   A new method of
mowing corn, grass, and other things, by means of
a machine moving on wheels, which may be
worked either by men or horses. PLUCKNET, T.J. For
another early patent see 2,877 (1805).
*First machine to incorporate a roller linked by
cogs to a spiral cutter*   5,990 (1830)   A new
combination and application of machinery for the
purpose of cropping or shearing the vegetable
surface of lawns, grass plats, and pleasure
grounds, constituting a machine which may be
used with advantage instead of a scythe for that
purpose. BUDDING, EDWARD

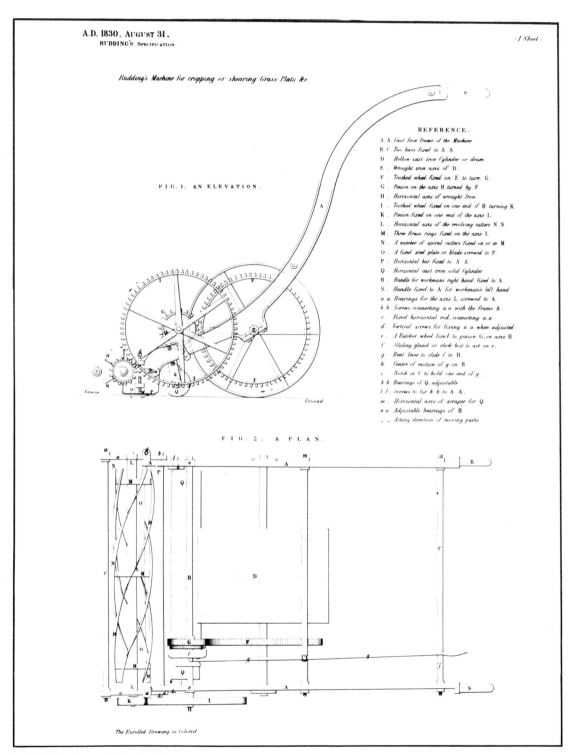

A.D. 1830, AUGUST 31.
BUDDING'S SPECIFICATION

(1 Sheet)

Budding's Machine for cropping or shearing Grass Plats &c.

FIG. 1. AN ELEVATION.

REFERENCE.

A A . Cast Iron frame of the Machine
B C . Two bars fixed to A A
D . Hollow cast iron Cylinder or drum
E . Wrought iron axis of D
F . Toothed wheel fixed on E to turn G
G . Pinion on the axis H turned by F
H . Horizontal axis of wrought iron
I . Toothed wheel fixed on one end of H turning K
K . Pinion fixed on one end of the axis L
L . Horizontal axis of the revolving cutters N N
M . Three Brass rings fixed on the axis L
N . A number of spiral cutters fixed on or in M
O . A fixed steel plate or blade screwed to P
P . Horizontal bar fixed to A A
Q . Horizontal cast iron solid Cylinder
R . Handle for workman's right hand fixed to A
S . Handle fixed to A for workman's left hand
a a . Bearings for the axis L screwed to A
b b . Screws connecting a a with the frame A
c . Fixed horizontal rod connecting a a
d . Vertical screws for fixing a a when adjusted
e . A Ratchet wheel fixed to pinion G, on axis H
f . Sliding gland or click box to act on e
g . Bent lever to slide f to H
h . Centre of motion of g on B
i . Notch in C to hold one end of g
k k . Bearings of Q adjustable
l l . Screws to fix k k to A A
m . Horizontal axis of scraper for Q
n n . Adjustable bearings of H
. _ Acting direction of moving parts

FIG. 2. A PLAN.

The Enrolled Drawing is Coloured

172 Budding's cog-wheeled cylinder cutter.

*Introduction of a plate to deflect grass cuttings
forward over the blades*   21,049/1900
Improvements in apparatus for cutting grass or
lawns. RANSOME, JAMES EDWARD

See also entry for ROTARY LAWN MOWERS.

173   Leather cloth
2,359/1855   Certain preparations of oils for and
solutions used when waterproofing and for the
manufacture of various articles by the use of such
compounds. PARKES, ALEXANDER

Leather cloth, consisting of a cotton cloth material
coated with a nitrocellulose solution and embossed
to give the appearance of grain leather, was first
manufactured under the name of Parkesine Co.
Ltd in 1866. It was intended as a cheap substitute
for hide but was not, at that time, commercially
successful. The industry was re-established in the
1890s but did not expand greatly until the
development of the motor industry created a
powerful demand for this type of cloth.

174   Linear induction motors
12,364/1895   Improvements in the shuttle
mechanism of reeds of looms. Justice, Philip
Middleton for THE WEAVER JACQUARD AND
ELECTRIC SHUTTLE COMPANY

The recent resurgence of interest in linear motors
in England has been led by ERIC ROBERTS
LAITHWAITE. He began working on the subject in
1947 at first applying himself to the high-speed
shuttle problem. He was successful in designing a
linear high-speed motor capable of producing
rapid reversals of direction as required for shuttles,
see 763,362 (1956) and 866,772 (1961). He later
turned his attention to other applications, notably
to those related to railway traction.

175   Linoleum
1,037 and 3,210/1863   Fabric for covering
floors, etc. WALTON, FREDERICK

176   Linotype
1,833/1885   Improvement in machines for
producing stereotype matrices and the like. Boult,

Alfred Julius for MERGENTHALER, OTTMAR. The
United States patent was 312,145 (1885).

177   Liquid crystals
*Nematics using domain alignment for light
modulation*   441,274 (1936)   Improvements in
or relating to light valves. LEVIN, BARNETT and
LEVIN, NYMAN, Marconi Wireless Telegraph
Company Ltd
Cholesteric liquid crystals   1,123,117 (1968)
Electric field device. WESTINGHOUSE ELECTRIC
CORPORATION. See fig. 2, p. 11.
Nematics using dynamic scattering   1,131,688
(1968)   Electro-optical elements and optical
display devices. WILLIAMS, RICHARD, Radio
Corporation of America
Liquid crystal with memory properties
1,246,847 (1971)   Liquid crystal display element
having storage. RADIO CORPORATION OF AMERICA

About one half per cent of all organic compounds
are liquid crystals in that within a limited
temperature range or within a given concentration
range of a solvent they possess a mesomorphic
phase between solid and liquid in which their
molecules arrange themselves in an ordered
pattern. There are three basic types of pattern
known as smectic, nematic and cholesteric which
are well described in patent 1,123,117. The optical
and other properties of these materials offer the
promise of a multitude of applications in the
electronics industry. Used as thin films they are
cheap and need very small amounts of power to
operate. Widespread research has been launched
to develop possible applications in such varied
fields as optical character recognition, random
access computer storage, colour television
screens, watch and calculator readouts, light
amplification, plotting radio frequency radiation
patterns, projection screens, integrated circuit
testing and crack detection. While the first of
the above patents was the first to give a detailed
description of the use of liquid crystals a passing
reference to their use for light modulation was
made in 358,087 (1931) LOISEAU, LOIS MARIE
JEAN, Compagnie pour la Fabrication des Comp-
teurs et Matériel d'Usines à Gaz.

178   Lithography

2,518 (1801)   For printing on paper, linen, cotton, woollen and other articles. SENEFELDER, JOHANN NEPOMUK ALOYS

179   Luminous paint

4,152/1877   Improvements in painting, varnishing, and whitewashing. BALMAIN, WILLIAM HENRY

180   Machine guns

418 (1718)   A portable gun or machine called a Defence, that discharges soe often and soe many bullets, and can be soe quickly loaden as renders it next to impossible to carry any ship by boarding. PUCKLE, JAMES

790/1865   Improvements in firearms. GATLING, RICHARD GORDON

3,493/1883   Improvements in machine or battery guns in cartridges for the same and other firearms. MAXIM, HIRAM STEVENS

181   Machine gun fire from aircraft

115,021 (1918)   Improvements relating to machine guns. SIEGLER, HENRI PHILIPPE ERNEST

This specification relates to means for controlling the fire of machine guns between the propeller blades of an aircraft by operating one, or more, guns from a shaft driven by the propeller motor.

182   Macintosh

4,804 (1823)   Process and manufacture for rendering the texture of hemp, flax, wool, cotton, silk, and also leather, paper and other substances, impervious to water and air. MACINTOSH, CHARLES

It was proposed that rubber, dissolved using coal oil which was becoming available in quantity as a by-product of the distillation of coal tar from pitch, should be used to cement two layers of fabric together in such a way that the user was protected from the stickiness of the rubber. This sandwich-type construction was not entirely new, as Spanish scientists working in Mexico had used the idea when developing leak-proof containers for mercury and CHARLES GREEN had adopted the principle in constructing a balloon envelope in 1821.

183   Magnetic sound recording

First practical system   8,961/1899   Method and apparatus for effecting the storing up of speech or signals by magnetically influencing magnetisable bodies. POULSEN, VALDEMAR

A.c. bias   United States 1,640,881 (1927)   Radio telegraph system. CARLSON, WENDELL L. and CARPENTER, GLENN W.

First proposal for the use of coated plastic tape   German 500,900 (1930) Lautschriftträger. PFLEUMER, FRITZ

High-frequency bias to oxide-coated tape   German 743,411 (1940)   Verfahren zur magnetischen Schallaufzeichnung. VON BRAUNMUHL, HANS JOACHIM and WEBER, WALTER

Noise reduction system   1,120,541 (1968) Improvements in noise reduction systems. DOLBY, RAY MILTON, Dolby Laboratories

The last of these patents describes the Dolby 'A' system for reducing tape hiss. A more recent revision of this method, called the 'B' system, is now being applied to domestic audio-cassettes.

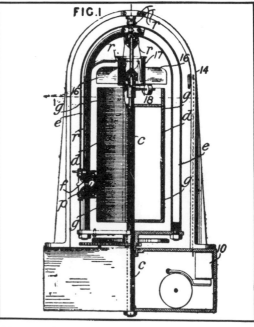

FIG.1

183.   In the form shown here this invention allows an electromagnet, on slider F, to be brought into contact with steel wire wound on to the cylinder. In an alternative form steel ribbon, mounted on two drums, is drawn across an electromagnet.

184   Magneto (for cars)
16,907/1898   Improvement in or relating to
ignition pins for explosion motors. BOSCH, ROBERT

This patent relates to an 'ignition pin' used with a
magneto ignition device.

185   Magnetron data storage
716,616 (1954)   Improvements relating to
registers such as are employed in digital
computing apparatus. REY, THOMAS JULIUS,
Electrical and Musical Industries Ltd

186   Maltbread
4,336/1886   Improvements in and connected
with the manufacture of bread and biscuits.
MONTGOMERIE, JOHN

187   Manipulators
781,465 (1957)   Improvements in or relating to
positioning, assembling or manipulating
apparatus. KENWARD, CYRIL WALTER

A description is given of robot hands which can be
applied to assembly work, for example, on car
production lines.

188   Margarine
2,157/1869   Producing fatty bodies. MEGE,
HIPPOLYTE

189   Masers and lasers
USSR 123,209 (1959)   A method for the
amplification of electromagnetic radiation
(Ultraviolet, visible, infrared and radio waves).
FABRIKANT, V. A.; VUDYNSKII, M. M. and BUTAYEVA, F.
(This patent was submitted as early as 18 June
1951 and is supplemented by USSR 148,441.)
United States 2,929,922 (1960)   Masers and
maser communications system. SCHAWLOW,
ARTHUR L. and TOWNES, CHARLES H., Bell Telephone
Laboratories Inc.

The principles of the process of stimulated
emission of radiation was first explained
theoretically by ALBERT EINSTEIN in 1916.
Fabrikant's Russian patent appears to be the first

suggestion in the patent literature to apply this
principle to a practical device. However, at the
time no suitable amplifying material was known
which would have enabled a practical device to
have been made. The first open publication of an
account of the maser principle was by J. WEBER in
June 1953 and the first description of a workable
maser was given in a paper by Townes and two of
his students in 1954. The second of the two patents
noted above claimed a 'practically realisable,
efficient, low-noise' laser. It called for power
sources not readily available at the time and there
were other difficulties. At about the same time
GOULD, GORDON filed a United States patent with
similar claims. British patents are 953,721 to
953,727 (1964), but there were difficulties in
practice. On 7 July 1960, T. H. MAIMAN made an
announcement to the press regarding what is now
considered to be the first practical laser; this notice
was soon afterwards published in *Nature*.
Maiman's laser used ruby as the amplifying
medium when most of the experts of the day
anticipated that the breakthrough to lasers would
come with the use of some gaseous medium.

190   Matches
5,732 (1828)   Producing instantaneous light.
JONES, SAMUEL

This patent relates to a match known as the
Promethean. A tiny, sealed vesicle of glass about
$\frac{3}{16}$ inch long, containing a drop of concentrated
sulphuric acid was surrounded by a mixture of
chlorate of potash, sugar and gum. This lit when
nipped with a small pair of pliers. The first to make
and sell friction matches was JOHN WALKER, a
chemist with a small shop at Stockton on Tees.
Walker did not patent and only made the matches
for about three years. Copies of his matches were
made and sold in London by Jones under the name
of *Lucifer*. Walker's original matches were made
of sticks dipped in a mixture of sulphide of
antimony and chlorate of potash. These were not
as efficient as the various types of matches
containing phosphorous which soon came onto
the market.

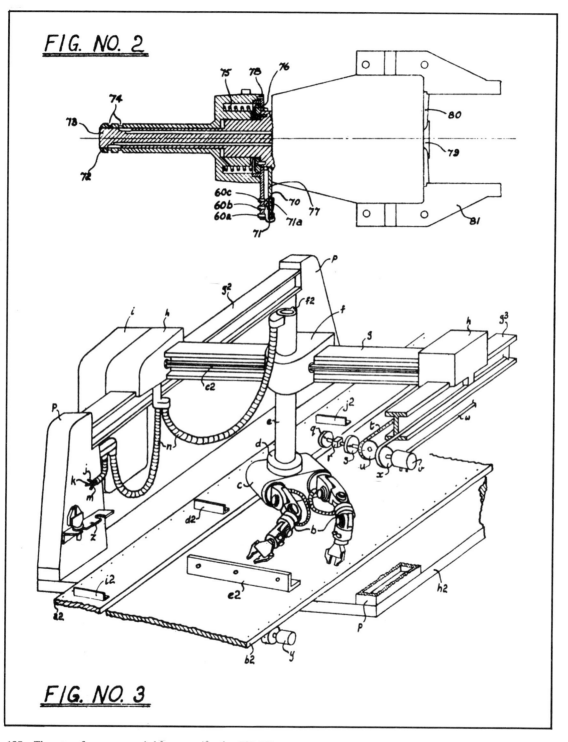

FIG. NO. 2

FIG. NO. 3

187 These two figures are copied from specification 781,465.

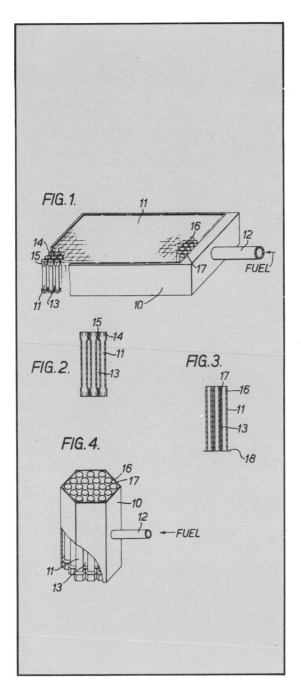

FIG. 1.

FIG. 2.

FIG. 3.

FIG. 4.

FUEL

191 The patentee writes that 'the burners described in this specification are compact, i.e. they provide good heat output unit area, they operate with a low pressure drop across the burner and they are capable of operating with a wide range of fuels and excess air conditions. Furthermore they can be made in a wide variety of shapes.'

191 Matrix burner
1,205,432 (1970)   Burner for use with fluid fuels. DESTY, DENIS HENRY and WHITEHEAD, DAVID MONTAGUE, British Petroleum Company Ltd

This new gas burner will burn natural or town gas with a silent flame and could lead to a drastic reduction in the size of domestic heating boilers and new types of radiant heater.

192 Measuring manufactured articles
805,391 (1958)   Improvements relating to apparatus for inspecting the dimensional accuracy of a workpiece. BOOTH, RICHARD HERBERT and PAYNE, EDWIN MALCOLM. Electrical and Musical Industries Ltd

A sensor passing over a manufactured piece produces an electrical signal which is compared with a signal indicative of what the correct shape should be.

193 Meat analogue
699,692 (1953)   Improvements in or relating to protein food products. BOYER, ROBERT ALLEN

This was the first patent for spinning filaments of protein foods, combining and flavouring them to yield product with the colour, flavour, taste and chewiness of real meat. Any type of animal or vegetable protein capable of being made into filaments might be used, for example, oil seed proteins such as soya bean or groundnuts, casein and other products obtained from gluten, fish or maize. The preparation of artificial meat from non-meat proteins had been proposed earlier, for example see United States 1,001,150 (1911), KELLOGG, JOHN HARVEY, but the idea of spinning filaments to simulate the texture was new.

194 Meccano set
587/1901   Improvements in toy or educational devices for children and young people. HORNBY, FRANK

195 Mechanical stoking
4,387 (1819)   Steam engines and furnaces of steam engines. BRUNTON, WILLIAM

This consisted of a circular rotating grate, coal was

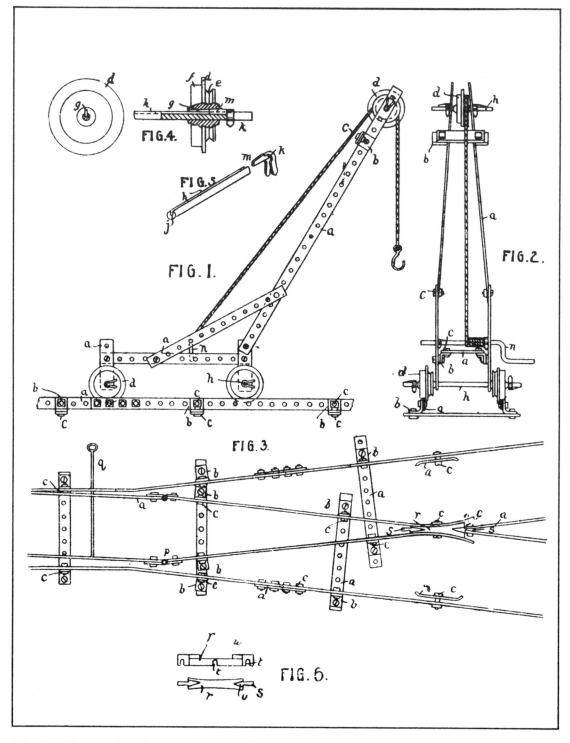

194 A model crane and set of points to be made from Hornby's Meccano.

fed continuously at a fixed point and was burnt as it travelled round, clinker being mechanically removed before the coal hopper supply. Later 4,685 (1812), Brunton patented a straight-line moving grate stoker. The sprinkling type stoker in which coal was sprinkled, or flicked, on to a stationary grate was patented in 1834 (patent 6,703, STANLEY, JOHN). JOHN GEORGE BODMER entered the field in 1834 (patent 6,617) with a suggestion for carrying the coal through a furnace in a cylindrical cage. This was not a success but later led Bodmer to suggest the moving chain-grate stoker patented later by JOHN JUCKES, 9,067 (1841).

196   Mercerising fabrics
14,024/1852   Preparation of fabrics for dyeing and printing. MERCER, JOHN and GREENWOOD, JOHN

197   Metal plating
5,127/1878   Plating metals. GLASER, FRIEDRICH

198   Metronome
3,996 (1815)   Metronome or musical timekeeper. MAEXEL, JOHN

199   Microrecording
316,668 (1929)   Process and devices for condensing on films photographic reproductions of printed or other works and for selectively reproducing from these films. SEBILLE, GEORGES

Earliest disclosure of projecting microimages used for recording books, documents, etc. The patent is concerned with a method of marking the film so that a particular page of the work recorded can be located quickly. See under STEREOSCOPIC CAMERA for a note on microphotography.

200   Microswitch
419,917 (1934)   Improvements in snap action electric switches. McGALL, PHILIP KENNETH, C.F. Burgess Laboratories, Inc. The equivalent United States patent was 1,960,020 (1934).

The switch described in the patent above was specially designed for use with a chicken brooder thermostat. It utilised the then newly discovered properties of heat-treated beryllium copper, which is an excellent spring material and good electrical conductor. The device was designed so that a very short travel of the switch plunger stores energy in a spring and used it to transfer the movable contact with a positive snap. The switch proved to have very wide applications and C. F. Burgess Laboratories formed a separate Electronic Division to develop the patent.

201   Milking machine
15,210/1889   Milking cows.
MURCHLAND, WILLIAM

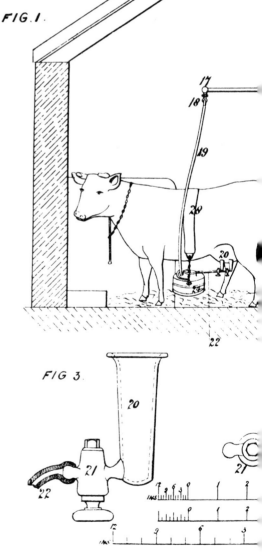

FIG.1.

FIG 3.

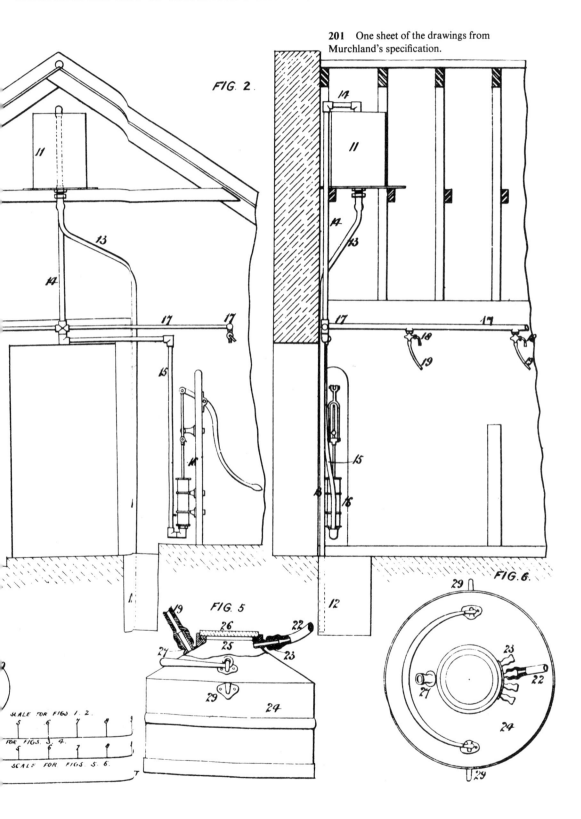

201 One sheet of the drawings from Murchland's specification.

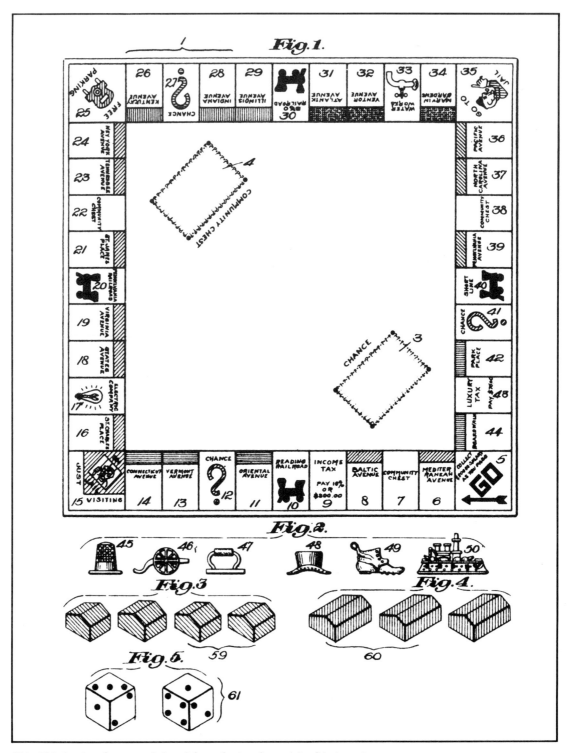

204 This drawing of a monopoly board shows the American origin of the invention.

202 Mirrors

9,968 (1843) Coating glass with silver, for looking glasses and other uses. DRAYTON, THOMAS

1,258–9 (1876) Improvements in apparatus for silvering mirrors, plate and other glass. PRATT, JAMES EDMONDSON

345,676 (1931) Improvements in or relating to the production of mirror glass. WALDRON, FREDERIC BARNES and HARRISON, ALLEN, Pilkington Brothers Ltd

The German Professor JUSTUS VON LIEBIG was the first to make a determined effort, between 1858 and 1864, to find a method of silvering glass by chemical means. He failed to produce a viable commercial process. Drayton's method, which was the first to be patented in England, was not highly successful and was followed by a number of variant proposals of which Pratt's was one of the more practical. The Pilkington patent describes a continuous silvering process.

203 Mirror galvanometer

329 (1858) Improvements in testing and working electric telegraphs. THOMSON, SIR WILLIAM (LORD KELVIN)

204 Monopoly

453,689 (1936) Improvements in or relating to apparatus for playing games. PARKER BROTHERS INC.

205 Monotype machine

8,183/1887 Improvements in the art of printing. LANSTON, TOLBERT. See also 7,399/1896 BOULT, ALFRED JULIUS.

206 Moving staircase

United States 25,076 (1859) Improvements in revolving stairs. AMES, NATHAN and McLEAN, WARD.
United States 470,918 (1892) Endless conveyor or elevator. RENO, JESSE W.
United States 479,864 (1892) Elevator. WHEELER, GEORGE A.

The Ames patent is the first known to propose the idea of a moving staircase. The last two patents noted above described practical schemes which were bought up by the OTIS firm who were able to combine the ideas in them to produce a working and saleable escalator.

207 Netlon

836,555 (1960) and 836,556 Improvements relating to the production of net or netlike fabrics by extrusion methods. MERCER, FRANK BRIAN, Plastic Textile Accessories Ltd

208 Newspaper printing presses

First cylinder press patent 1,748 (1790) Machine for printing on paper, linen, cotton, woollen and other articles. NICHOLSON, WILLIAM
First practical steam powered machine 3,321 (1810) Printing by machinery. KOENIG, FREDERICK. See also 3,496 (1811), 3,725 (1813) and 3,868 (1814).
First successful machine to use stereotype plates 4,194 (1818) Printing presses. COWPER, EDWARD
First rotary pres to be used in England 11,505 (1846) Machines for printing papers and other fabrics. APPLEGARTH, AUGUSTUS

The Nicholson press foreshadowed many of the features of the later presses but failed for want of financial support. Koenig presses were introduced to print *The Times* in November 1814 replacing the old Stanhope hand presses. The Applegarth press replaced the Koenig presses in 1848 but were preceded in the United States by machines with many similar features designed by RICHARD M. HOE. Hoe took out a number of United States patents, see for example 5,199 (1847). His machines were used to print the *Philadelphia Ledger* in 1847. A somewhat more primitive machine had been used to print the *Christian Advocate* in New York in 1828 and a patent for a rotary press had been taken out in England, 6,762 (1835), HILL, ROWLAND. These machines were all sheet-fed. The first web-fed machine was introduced by WILLIAM BULLOCK in New York, United States 38,200 (1863). Bullock died from injuries sustained when he was trapped by this machine four years later. Applegarth patented in England web-fed machines

A.D. 1883, 31*st December.* Nº 5978.

Manufacture of Carbons for Incandescent Electric Lamps.

LETTERS PATENT to Joseph Wilson Swan of Bromley in the County of Kent Chemist for an Invention of 'IMPROVEMENTS IN THE MANUFACTURE OF CARBONS FOR INCANDESCENT ELECTRIC LAMPS'

PROVISIONAL SPECIFICATION left by the said Joseph Wilson Swan at the Office of the Commissioners of Patents on the 31st December 1883.

JOSEPH WILSON SWAN, of Bromley in the County of Kent, Chemist, "IMPROVEMENTS IN THE MANUFACTURE OF CARBONS FOR INCANDESCENT ELECTRIC
5 LAMPS"

It is necessary for obtaining the greatest efficiency in electrical lamps in which light is produced by the incandescence of a carbon filament that the filament should be of a homogeneous and solid texture and that its section should be uniform in size.
10 Hitherto there has been experienced great difficulty in producing thin carbon filaments of a perfectly uniform sectional size, and at the same time homogeneous and solid in texture.
My Invention has for its object the production of carbon filaments possessing these qualities and the removal of the difficulties which have hitherto stood in the
15 way of their production.
According to my said Invention I take a carbonisable material in a plastic or semi-liquid state and I squeeze or press it through a hole or die so as to form a thread or threads of equal substance and of any required length and also of any required form in transverse section—The material I may employ in carrying out
20 my Invention is a solution of Nitrocellulose in Acetic Acid, with or without the addition of other well known solvents of Nitrocellulose and with or without other carbonisable material than nitrocellulose. I do not limit myself to this particular solution of Nitrocellulose nor to the use of a Nitrocellulose solution at all as the carbonisable material, for my process of forming carbonisable thread by squeezing a
25 carbon liquid or paste through a die is applicable to any suitable carbonisable material in a plastic, liquid, or semi-liquid condition, for example nitro-glucose rendered plastic by heat may be employed—In employing an Acetic Acid Solution of Nitrocellulose I project the filament as it issues from the die into water or other liquid which has the effect of instantly setting or giving coherence to the issuing thread.   5
When the thread has been formed in the manner described I treat it with a solution of Hydro-sulphate of Ammonia or other equivalent deoxidising agent until it is no longer in a condition to burn explosively, and I then dry it and shape it into the forms in which it is to be used in the lamps and I carbonize it in any suitable manner.   10

212   This is a reproduction in full of Swan's provisional specification, which runs to just over a page in length. Few modern specifications are this short, a typical length would be from three to five pages although some run to a considerable length when the invention involves complicated machinery or there are several 'examples' or 'embodiments' described.

that also used stereotype plates and included cutting and folding equipment, 2,098/1859. The 'Walter' presses introduced to *The Times* in 1868 were developments of machines of this type and delivered completed newspapers, cut and folded, at the rate of 12,000 copies per hour.

209 Nickel plating
8,407 (1840)   Preserving and covering certain metals and alloys of metals. SHORE, JOSEPH

This was the first patent for a commercial nickel plating process. The use of a neutral ammonium sulphate bath was later proposed in a paper by G. GORE in 1855. I. ADAMS developed this latter process to commercial use, patenting it in the United States, 52,271 (1866), and applying it to the plating of gas burner tips. Adams later developed the technique of using nickel ammonium sulphate baths in United States 93,157 (1869). This process gave him a virtual monopoly of nickel plating in the United States for many years.

210 Nitric acid
698/1902   Improvements in the manufacture of nitric acid and nitrogen oxides. OSTWALD, WILHELM

211 Nitrocellulose
283/1855   Obtaining and treating vegetable fibres. AUDEMARS, GEORGE

212 Nitrocellulose thread for lamp filaments
5,978/1883   Improvements in the manufacture of carbons for incandescent electric lamps. SWAN, JOSEPH WILSON

213 Nozzle looms
773,531 (1957)   Improvements in looms for weaving. MIRA, ZAVODY NA PLETENE A STAVKOVE ZBOZI NARODNI PODNIK (Czechoslovakia)

A new development in weaving in which the weft thread is thrown across the shed of the loom by high-pressure jet of water.

214 Nuclear fission reactors
614,156 (1948)   Apparatus for the production of energy by nuclear fission. CENTRE NATIONAL DE LA RECHERCHE SCIENTIFIQUE. Fig. p. 92.

Although this has a publication date of 1949 it was first applied for in England in 1940, and has a priority date of 1 May 1939 when application was made in France. It was the first patent to propose a method of producing energy by nuclear fission.

215 Nylon
461,236–7 (1937)   The manufacture of new compositions of matter and of artificial filaments, fibres, sheets, films and the like therefrom. E. I. DU PONT DE NEMOURS AND CO.

216 Ocarina
1,020/1877   Musical wind instrument. LAKE, WILLIAM ROBERT

217 Orlon
585,367 (1946)   Production of an improved acronitrile polymer yarn. E. I. DU PONT DE NEMOURS AND COMPANY. See also the companion patent 585,368 (1946).
944,424 (1963)   Improvements relating to crimped composite filaments of synthetic polymers. E. I. DU PONT DE NEMOURS AND COMPANY

The first of the patents above is one of a series on acronitrile polymers taken out shortly after their discovery. The second patent describes a later development in which a dry-spinning process employing a pair of acronitrile polymers is used to produce, by chemical means, a 'crimp reversibility' giving a fibre much closer in structure to natural wool. See fig. p. 93.

218 Paper making
Fourdrinier machines   2,951 (1806)   The method of making a machine for manufacturing paper of indefinite length, laid and wove with separate molds, FOURDRINIER, HENRY
Cylinder type machines   3,191 (1809)   Certain improvements on my former patent machinery for

PATENT SPECIFICATION

Application Date: March 12, 1934. No. 7840/34.

„   „   July 4, 1934.   No. 33540/35.

(No. 33540/35 being divided out of Application No. 19721/34.)

Application Date: Sept. 20, 1934.   No. 27050/34.

One Complete Specification Left: April 9, 1935.

(Under Section 16 of the Patents and Designs Acts, 1907 to 1932.)

Specification Accepted: Dec. 12, 1935.

440,023

PROVISIONAL SPECIFICATION
No. 7840 A.D. 1934.

Improvements in or relating to the Transmutation of Chemical Elements

I, Leo Szilard, a citizen of Germany and Hungary, c/o Claremont Haynes & Co., Vernon House, Bloomsbury, Square, London, W.C.1, do hereby declare the nature of this invention to be as follows:—

It has been demonstrated that if atoms or nuclei, e.g. hydrogen atoms (or protons), heavy hydrogen atoms, referred to from now onwards as diplogen, (or diplogen ions, referred to from now onwards as diplons) etc. are shot at chemical elements, a definite fraction of these shooting particles will cause transmutation in many elements. (How large this fraction is will depend on the nature of the element, the nature of the shooting particle, and its velocity.) If one uses the above mentioned particles and shoots them on light or heavy hydrogen lithium (6) or lithium (7) or other elements a certain proportion of the particles lose their energy through ionizing the substance through which they are shot, and only a fraction of the shooting particles will meet a nucleus of the substance before losing so much energy that the shooting particle is unable to cause transmutation in nuclei which it meets. Of these particles which meet a nucleus in their path (while still being in possession of a sufficiently large fraction of their initial energy) again only a further fraction will be able to penetrate the nucleus, (will be able to cause a transmutation); if the shooting particles are positively charged they are repulsed by the positively charged nucleus, and the probability of their penetrating the nucleus is a function of their relative velocity.

This probability rises rapidly with increasing velocity of the shooting particle and eventually reaches unity at a velocity which depends both on the nature of the shooting particle and the nature of the bombarded element.

However, even if this probability is equal to unity one still has to face the fact that a shooting particle has to travel for instance in air a large distance in order to encounter a nuclear collision (which may cause transmutation), but due to the energy loss which it suffers through ionizing the air its range is comparatively small if its initial velocity corresponds to several million volts energy. Only a fraction of the above mentioned shooting particles can therefore produce transmutation if shot into air or other substances or similar characteristics concerning ionization losses and nuclear collisions.

In accordance with the present invention radio-active bodies are generated by bombarding suitable elements with neutrons, which can be produced in various ways.

In accordance with one feature of the present invention nuclear transmutation leading to the liberation of neutrons and of energy may be brought about by heating up a small area filled with suitable elements very suddenly to high temperature by means of an electric discharge.

Radio Active Substances.

It is possible to produce elements capable of spontaneous transmutation by bombarding certain elements with fast charged nuclei, for instance by bombarding carbon with protons or aluminium, boron and magnesium with helium ions (particles). However, most of the radio active elements produced by the bombardment of these light elements with protons or alpha particles have a short existence (they disintegrate spontaneously in a time shorter than a few hours to half their amount), and it is not possible to use these charged nuclei for the transmutation of the heavier elements with good efficiency as the ionization loss gets too large. It is, however, possible to produce with good efficiency (both from lighter and heavier elements) radio active sub-

[Price 1/-]

---

214   The patent listed under the heading NUCLEAR FISSION REACTORS indicates that the basic proposals for nuclear power had been made before the start of the Second World War. Proposals for the production of radioactive substances had in fact been patented several years before, as can be seen from the patent illustrated here. Jacob Bronowski, in a radio talk in 1975 (*The Listener*, May 29, 1975, p. 701), suggests that Tsilard had written a specification for an atomic bomb as early as 1933 and presented it secretly to the British Admiralty.

944,424   5

inner annular space 7 and thence through the screen 4 and orifice 3 to form a part of a composite filament, while the second passes through the lead hole 10 to the outer annular space 6 and thence through the screen 4 and the outer side of the orifice 3 to form the other part of the composite filament.

The expression " intrinsic viscosity " with the symbol (n) as used herein signifies the value of ln(n), at the ordinate axis intercept i.e. when c equals 0) in a graph of

$$\frac{ln(n)_r}{c}$$

as ordinate with c values (grams per 100 ml. of solution) as abscissae. $(n)_r$ is a symbol for relative viscosity, which is the ratio of the respective flow times in a viscosimeter of a polymer solution and the solvent thereof. ln is the logarithm to the base e. All measurements in the following Examples on polymers containing acrylonitrile combined in the polymer molecule were made with DMF solutions at 25°C. The apparatus referred to is that described above.

EXAMPLE I.

This Example shows the preferred process and product of this invention.

A 20% solution in DMF of polyacrylonitrile of (n) 1.95 and containing 27 milliequivalents of acid groups per kilogram of polymer (as determined by titration in a DMF solution) (hereinafter designated polymer A) was fed to the annular chamber 9 at a rate of 47 grams/minute, and thence to the annular space 7 and out into the spinning cell which was 9 inches in diameter and 14 feet long as part of a composite filament so that it faced the centre of the spinning cell. Simultaneously, a 27% solution in DMF of a copolymer of acrylonitrile (96% by weight) and styrenesulphonic acid (4% by weight) of (n) 1.54 and analysing 240 milliequivalents of acid per kilogram of polymer (hereafter designated polymer B) fed to the annular space 8 at the rate of 47 grams/minute and then through the annular space 6, to be extruded as the other component of a composite filament so that it faced the wall of the spinning cell. The spinneret contained 140 orifices of

0.069 inch in diameter located on a 5.27 inch diameter circle. A mixture of carbon dioxide and nitrogen gases was circulated through the cell at a rate of 57 lb. per hour. Temperatures of 105°, 315° and 180°C. were used for the spinning solutions, head and cell respectively. The threadline consisting of 140 composite filaments was wound up at 200 ypm and contained 19% of DMF and 1.0% water based on its dry weight.

Three hundred and fifty ends of yarns produced as above with a combined denier of 382,000 were combined into a tow and drawn to 4 times their original length (i.e. 4X draw ratio) in baths of water at 95°C. which extracted the residual DMF. The drawn and unrelaxed wet tow was mechanically crimped in a stuffer box to an extent of 6—7 herringbone crimps per extended inch using a stuffer box temperature of 50°C. The crimped tow was then cut into 3 1/8" length staple. The cut staple, loosely arranged in a tray, was dried for 15 minutes in a circulating air oven at 270—275°F. The dried staple had a weak mechanical crimp of 6 to 7 crimps per inch plus 6—8 helical crimps per inch.

The above-prepared staple develops 16 helical crimps per inch of extended length when boiled free of restraint in water. The helically crimped fibres thus prepared have an ECR of 40% and a Δcpi of 6.4 crimps per inch. Substantially all of the mechanical crimp is removed. The staple has a tenacity of 2.2 grams per denier (gpd) and an elongation at the break of 40% and a denier per filament of 3.0 (0.34 Tex) after boiling and drying.

EXAMPLE II.

The properties of the product obtained in Example I are compared with the properties of other composite fibres in Table I below. It will be noted that the fibre of Example I meets the requirements of the present invention since it has a high crimp reversibility (ECR of 25% or larger) coupled with a low crimp intensity (10 to 18 crimps per inch of extended length). Data listed in Table I for other fibres not made by this process show that a reduction in crimp is accompanied by a reduction in crimp reversibility.

217 A page of the patent showing how to put the crimp into Orlon fibres. This is reproduced here to illustrate the amount of detailed information that is contained in patent specifications.

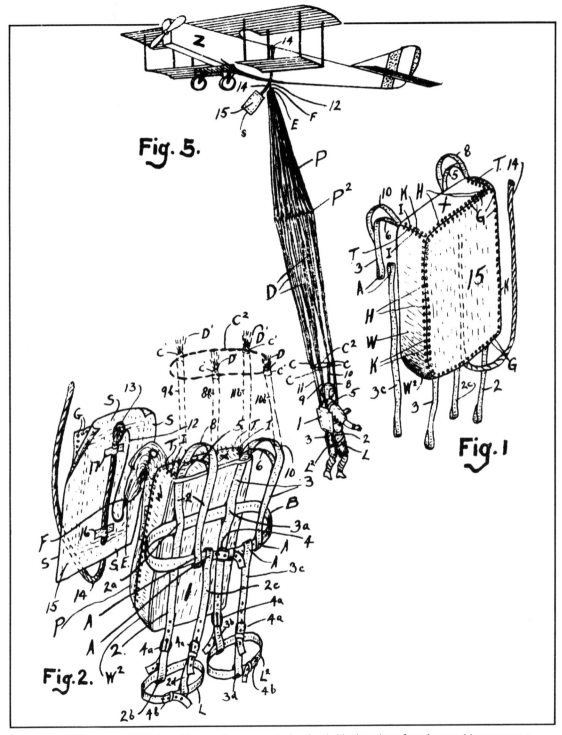

Fig. 5.

Fig. 1.

Fig. 2.

220   To quote from the specification—'In operation, upon entering the airship the aviator fixes the rope 14 to some part thereof near where he sits when operating his machine. When he wants to jump out, he does so and the rope 14, by its attachment to the cover 15, rips the cover off at the cords or threads which hold it in place by the eyelet G and H . . .'.

cutting and placing paper and also certain machinery for the manufacture of paper by a new method. DICKINSON, JOHN

Earlier paper making machines were developed by the Frenchman LEGER DIDOT and Englishman BRYAN DONKIN. Patents for these 2,487 (1801) and 2,708 (1803), both in the same name of JOHN GAMBLE, Didot's brother-in-law, were bought up by Fourdrinier and his brother. The main principles of the Fourdrinier machine are still in use today. The Dickinson patent is the first known British patent for a cylinder type machine although it is predated by a United States patent granted to CHARLES KINSEY in 1807.

219 Papier mâché
1,027 (1772) Making japanned high varnished panels in paper, for carriages and sedan chairs, also for rooms, doors, and cabins of ships, as well as for cabinets, screens, chimney pieces, tables, trays, caddies, tea-chests and dressing boxes. CLAY, HENRY
See also 2,830 (1805), JONES, THOMAS and 7,049 (1836) with 10,653 (1845) both BRINDLEY, WILLIAM

220 Parachute
138,059 (1920) Improvements in safety parachute pack device. IRVING, LESLIE LEROY, Irving Airchute Co.

This was the first to incorporate the ripcord.

221 Paraquat and diquat
813,531 (1959) New herbicidal compositions. BRIAN, ROBERT COLES; DRIVER, GEORGE WILLIAM; HOMER, RONALD FREDERICK and JONES, RICHARD LEWIS, Imperial Chemical Industries, Ltd

The first weedkiller which was capable of killing green vegetable matter when sprayed in solution on it, but yet was inactivated on contact with the soil.

222 Pen nib (steel)
3,118 (1808) Pens. DONKIN, BRYAN

This is the first British patent for a steel nib although there is evidence to show that steel

writing pens were made in Aix-la-Chapelle in 1784. Metal pens date much earlier, one having been found in excavations at Pompeii. Steel pens did not come into widespread use in England until about 1830.

223 Penicillin
552,619 (1943) Improvements relating to bactericidal substances. BOOTS PURE DRUG CO., LTD, and others.

SIR ALEXANDER FLEMING first discovered penicillin as early as 1928. He found it unstable and difficult to work with, and abandoned work on finding a technique for isolating it. A laboratory method of isolating and storing it was discovered by a team at Oxford, working under SIR HOWARD FLOREY, between 1939 and 1941. Commercial techniques for its large-scale production were later developed in the United States with government support.

224 Perspex (polymerised methyl methacrylate)
395,687 (1933) Manufacture of a new polymerisation product and moulded bodies therefrom. HILL, ROWLAND. Imperial Chemical Industries

225 Petroleum cracking
10,277/1889 Distilling mineral oils. DEWAR, JAMES and REDWOOD, BOVERTON
477,846 (1938) Process of and apparatus for the treatment or catalysis of hydrocarbons or other fluids for heat exchange. HOUDRY PROCESS CORPORATION

The first of these two patents relates to a batch thermal cracking method developed from an existing process to obtain kerosene. The kerosene process was carried out at atmospheric pressure, by increasing the pressure it was found that petrol could be obtained instead. The process was first used commercially by INDIANA STANDARD under the name of the Burton Process after WILLIAM M. BURTON who made some improvements, United States 1,049,667 (1913). The Burton process could not be operated continuously and several continuous thermal processes were developed and brought into operation. The Holmes–Manley process, which was based on an invention of JOSEPH H. ADAMS, United States 976,975 (1910), and developed by R. C. HOLMES and F. T. MANLEY was operated commercially from 1920. The Dubbs process, named after its inventors JESSE A. and CARBON B. DUBBS, was introduced to commercial use in 1922. The Houdry process, the subject of the second patent above was the first practical catalytic cracking process. The French inventor Eugene Houdry commenced work on the nature of catalysis and its effects on the cracking process as soon as the First World War had come to an end. He was first able to produce petrol suitable for internal combustion engines in 1927 but was not able to overcome the difficulties over the design of equipment for commercial scale use until 1936. At first the Houdry process was discontinuous but since continuous catalytic processes have been introduced, Thermofor Catalytic Cracking was introduced in 1944 and the Houdriflow process in 1950. Both were developed by research staff of SOCONY MOBIL OIL.

226 Phonograph

Cylinder machine 2,909/1877 Improvements in instruments for controlling by sound the transmission of electric currents and for the reproduction of corresponding sounds at a distance. EDISON, THOMAS ALVA

Disc type of machine 1,644/1878 Improvements in means for recording sounds, and in reproducing such sounds from such record. EDISON, THOMAS ALVA

The first of these was also an important patent in the development of the telephone, see under the entry for the telephone. It was filed before the first US phonography patent, 200,521 (1878). Edision's early machines recorded on to tin foil. Chichester A. Bell and Sumner Tainter, in US 341,214 (1886), proposed that the tin-foil be replaced by wax surface. The wave profiles were still being cut in a plane vertical to the recording surface so that there were distortions on playback since the needle met greater resistance when cutting deeper into the groove. Emile Berliner,

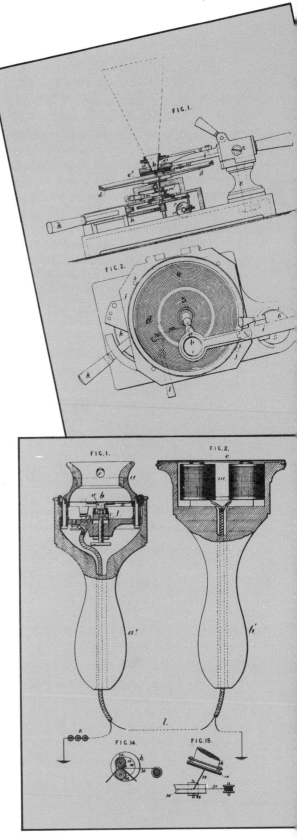

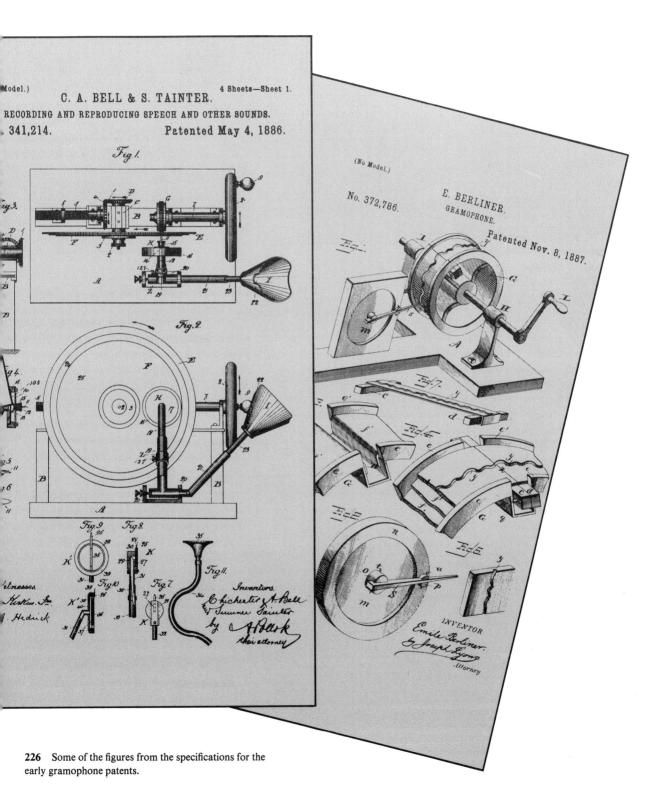

226 Some of the figures from the specifications for the early gramophone patents.

in US 372,786 (1887), proposed that the wave profile be cut horizontally on a thin sheet of wax, these wave profiles could then be etched on to a surface of harder material for playback. This latter technique showed the way towards the mass-production of records and also removed the distortion problems.

227 Photography

Calotype 8,842 (1841) Obtaining pictures or representations of objects. TALBOT, WILLIAM HENRY FOX

Daguerreotype 8,194 (1839) A new or improved method of obtaining the spontaneous reproduction of all the images received in the focus of the camera obscura. Berry, Miles for DAGUERRE, LOUIS JACQUES MANDE and NIÉPCÉ, JOSEPH NICEPH

The calotype process was a negative-positive process using a paper sensitised with silver iodide, developed with a gallic-acid, silver nitrate solution and fixed with hyposulphate of soda (see 9,753 of 1843). The daguerreotype was a positive process in which a copper sheet carrying a coating of metallic silver was sensitised by exposure to iodine vapour and after exposure was fixed with sodium thiosulfate. The daguerreotype was not patented in France but the government made financial recompense and offered the invention to the world. The daguerreotype was an immediate success but eventually gave way to the more convenient negative-positive technique.

228 Photo-typesetting

496,886 (1938) Improvements in and relating to the method of photographically producing type copy for printing processes. INTERTYPE LTD

229 Piano

English double action 1,571 (1786) Pianofortes and harpsichords. BEIB, JOHN

Upright piano 2,591 (1802) Certain new improvements in the construction and action of upright pianofortes. LOUD, THOMAS

First piano pedal 1,379 (1783) Pianofortes. BROADWOOD, JOHN

The English double action mechanism replaced the older and simple 'single action' for the better class pianos of the day. The 'single action' mechanism was never patented in England. The Loud upright piano was the first to be constructed of reasonably compact size.

230 Polariser

412,179 (1934) Polariser and method of making the same. LAND, EDWIN HERBERT

Methods of polarising light date from the Nicol prism of 1828. Methods of polarising car headlights had been devised before this patent and other work on polarising materials was in progress. Land's method involved forming many tiny crystals, orienting them all in the same direction and embedding them in a transparent covering.

231 Polaroid camera

658,740 (1951) Improvements in or relating to sheet material for use in photography. A. H. Stevens for LAND, EDWIN, International Polaroid Corporation

This is the first patent concerned with the Land or Polaroid type of camera which enables a photographic record to be obtained within a few seconds of exposure.

232 Polyimides

898,651 (1962) Polymers containing polyimide-acid recurring units and their production. EDWARDS, WALTER MURRAY, E. I. Du Pont de Nemours

This is not the first polyimide patent but was the first directed to the polyimide-acid intermediates which rendered polyimides commercially viable and since which intense development has taken place.

233 Polycarbonates

772,627 (1957) Process for the manufacture of thermoplastic polycarbonates. FARBENFABRIKEN BAYER AKTIENGESELLSCHAFT

The polycarbonates are an important class of

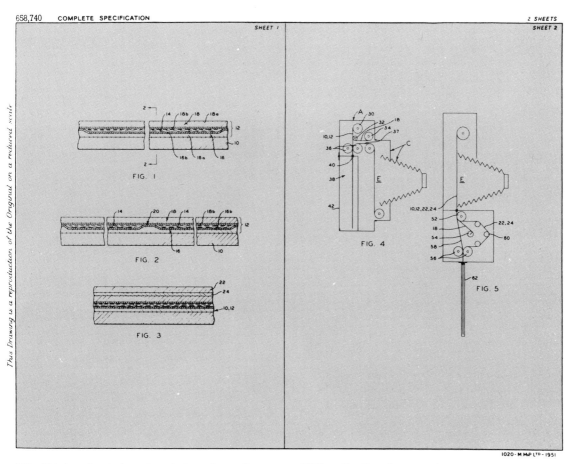

658,740   COMPLETE SPECIFICATION

This Drawing is a reproduction of the Original on a reduced scale

2 SHEETS

SHEET 1

SHEET 2

FIG. 1

FIG. 2

FIG. 3

FIG. 4

FIG. 5

1020-M M&P LTD-1951

231   This invention took amateur photography another leap forward by making the instant print a reality.

polyester plastics with a wide range of uses. This patent specification was the first to describe a method of preparing polycarbonates from di-(man-hydroxy-henyl)-substituted aliphatic hydrocarbons.

234   Polyethylene

471,590 (1937)   Improvements in or relating to polymerisation of ethylene. FAWCETT, ERIC and others, Imperial Chemical Industries

This process was the culmination of a decade or so of work by the Alkali Division of ICI. Later a new process which did not require the high pressures and temperatures implicit in this process was developed, 713,081 (1954), ZIEGLER, KARL.

235   Portland cement

1,135/1852   Portland and other cements. ASPDIN, WILLIAM

There is some controversy as to whether Aspdin was the true originator of Portland cement. Certainly his use of the term 'Portland' caught on and became generally used from his time, but there is even doubt whether he was the first to coin the term. JOHN SMEATON may well have used the word some years earlier during his experiments with 'water cements' in 1756, when constructing the Eddystone Lighthouse. Smeaton was experimenting with natural limestone materials. The next significant improvements were the 'Roman' cements introduced by JAMES PARKER, 2,120 (1796). Before the time of Aspdin's patent JAMES FROST was manufacturing a product not greatly different from Aspdin's under the name of 'British Cement', 4,679 (1822). Neither of these was as close in composition to presentday cements as was that being manufactured by WHITE AND SONS of Swanscombe and developed by ISAAC CHARLES JOHNSON. Mass production did not come until the 'bottle' kilns used by the early manufacturers was replaced by the rotary kilns invented by THOMAS RUSSELL CRAMPTON, 2,438/1877 and improved by FREDERICK RANSOME, 5,442/1885.

236   Post Office Position Indicator (POPI)
579,346 (1946)   An improved radio-aid to navigation. MITCHELL, HENRY THOMAS and KILVINGTON, THOMAS

237   Postal franking machines
13,360/1905   Coin freed franking apparatus. MOSS, ERNEST. See also 11,364/1911 and 106,283 (1917).

The first patent for a stamp-cancelling machine was 2,261/1863. Moss, whose patents are noted above, was a New Zealander and can claim the distinction of being the first to have a franking machine put to official use. He made several versions of his machine before producing a model that was approved by the New Zealand Post Office. The first franking machine operated in England was a penny-in-the-slot machine installed for public use in 1912. Its inventor was WILKINSON, FREDERICK, see patents 11,534/1911 and 17,707/1911. It was withdrawn because of the difficulty of preventing forgeries. The first machine to come into use in a commercial office in England was put to use by the Prudential Insurance Company on 5 September 1922. This was an American machine developed mainly by ARTHUR H. PITNEY whose first British patent was 21,234/1902.

238   Powdered milk
First process to give a marketable product
2,430/1855   Treating milk in order to preserve it. GRIMWADE, THOMAS SHIPP
An early drum-drying process   United States 712,545 (1902)   Process of preserving milk in dry form. JUST, JOHN AUGUSTUS
Principle of agglomeration, instantising

United States 2,835,586 (1958)   Dried milk product and method of making same. PEEBLES, DAVID DART.
See also United States 2,054,441 (1936) and 2,710,808 (1955).

239   Power steering
257,338 (1926)   Improvements in or relating to hydraulic power transmission gear. VICKERS LTD and THE VARIABLE SPEED GEAR LTD

240   Prestressed concrete
338,864 (1930)   Process for the manufacture of articles of reinforced concrete. FREYSSINET, EUGÈNE and SEAILLES, JEAN. See also 338,934 (1930).

There were many suggestions for prestressing before this time the earliest of which was probably that of P. H. JACKSON, United States patent in 1886. Freyssinet, however, was the first to use high strength wires so that sufficiently high stresses could be applied to make the process of practical value.

241   Process control
620,261 (1949)   Improvements relating to electronic timing apparatus. SHAND, GEORGE, Metropolitan-Vickers Electrical Company Ltd

The use of electronic timing apparatus to control a process involving a number of separate operations continuing in sequence each for different time periods. For a more advanced method of process control involving the use of computers and sampling see 997,049 (1965), E. I. DU PONT DE NEMOURS AND COMPANY.

242   Projection lamps
807,137 (1959)   Improvements in electric incandescent lamps. GENERAL ELECTRIC COMPANY

The lamps have a tungsten filament and a gas filling of iodine, or a mixture of inert gas and iodine. The iodine combines with evaporated metallic tungsten which dissociates when close to the filament thus redepositing the tungsten on the filament. Thus the lamps have a high luminous efficiency and a long life.

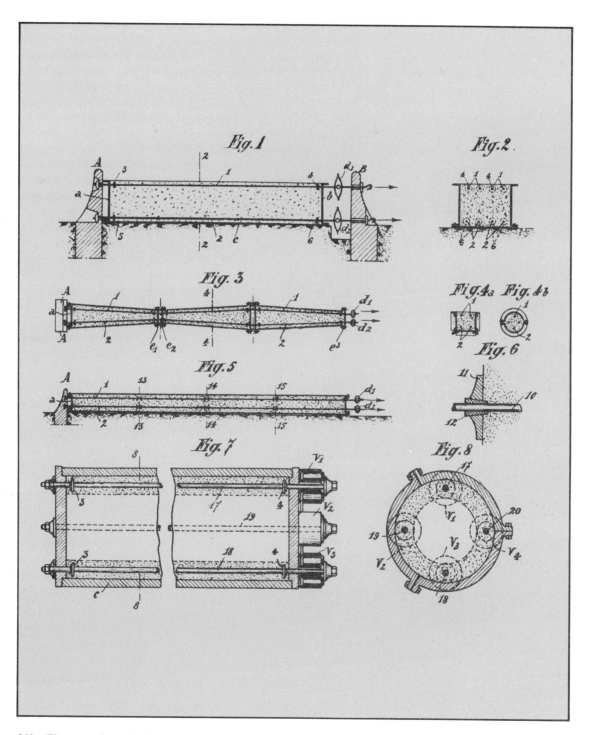

240 'The present invention has for its object a method of manufacturing articles of reinforced concrete moulded in advance and adapted to be employed or put in position only after setting and hardening has taken place, such articles being in the form of posts, girders, plates, railway sleepers, gutters, fences, panels or the like.'

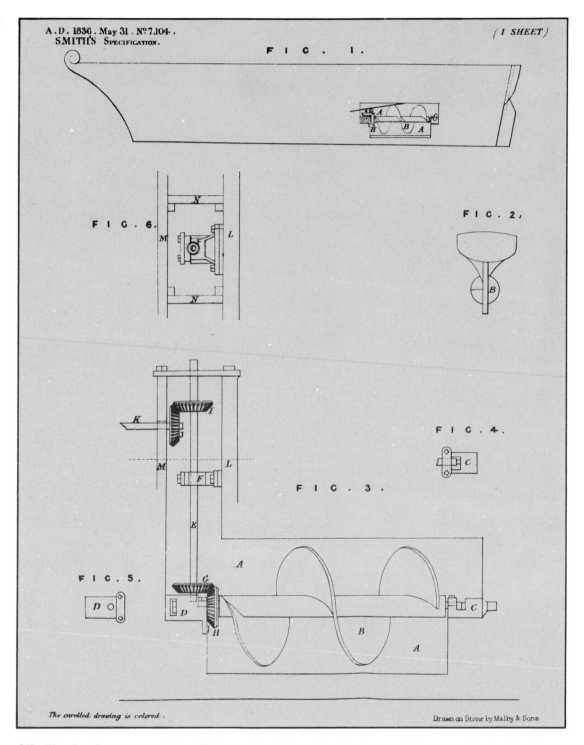

243   There have been many contenders claiming credit for the invention of the first practical screw propeller. Petit Smith's claims are at least as good as those of his contemporaries.

243 Propellers (screw type)

7,104 (1836) Propeller for steam and other vessels. SMITH, FRANCIS PETIT. For earlier patent see 2,000 (1794), LYTTLETON, WILLIAM.

244 Protamin-insulin

456,101 (1936) Improvements in and relating to insulin preparation. HAGEDORN, NORMAN JENSEN and WODSTRUP-NIELSEN, INGRID

245 Puddling

1,420 (1784) Shingling, welding, and manufacturing iron and steel into bars, plates and rods of purer quality and larger quantity, than heretofore, by a more effectual application of fire and machinery. CORT, HENRY

This method of 'dry puddling' followed on from Cort's developments on rolling (see ROLLING OF METALS). Like the patent for rolling the basic ideas incorporated were well known in the trade but Cort was able to combine these ideas into a commercially workable system which marked a significant advance in the practice of manufacture of wrought iron. Earlier puddling patents which met with less success were 851 (1766), CRANAGE, THOMAS and CRANAGE, GEORGE; and 1,307 (1782), ONIONS, PETER.

246 Puffed cereals

13,353/1902 Improvements in or relating to the treatment of starch and materials containing starch. Thompson, William Phillips for ANDERSON, ALEXANDER PIERCE

This describes the results of experiments conducted by Anderson in Minneapolis. The firm of QUAKER OATS took up the financing of further studies and Anderson moved to their works in Chicago. Puffed rice was introduced to the public at the World Fair in St Louis in 1904. At first it was sold as confectionery, keeping company with popped corn, but it was not long before it began to be marketed as a breakfast food.

247 Pulse code modulation

535,860 (1941) Electrical signalling systems. Standard Telephones and Cables Limited for LE MATÉRIEL TÉLÉPHONIQUE SOCIÉTÉ ANONYME

For an earlier French patent see 837,921 (1937). The technique substantially reduces the background noise level for transmissions of complex wave forms, e.g. speech transmissions.

248 Radar

593,017 (1947) Improvements in or relating to wireless systems. WATSON-WATT, ROBERT ALEXANDER

This was the first workable radar system. The basic idea was set out in an earlier Watson-Watt patent 591,130 (1947). Another patent of interest is 520,778 (1940), Standard Telephones and Cables Limited for LE MATÉRIEL TÉLÉPHONIQUE

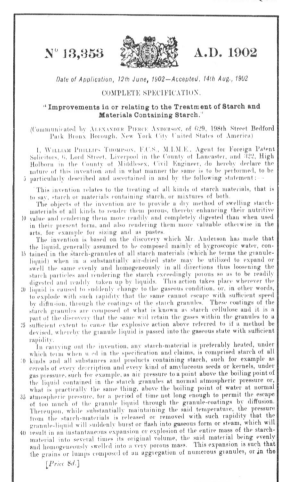

246 The first page of the specification of an invention that changed the breakfast habits of so many.

SOCIÉTÉ ANONYME. This, although published before 593,017, has a later priority date since the publication of the latter was delayed. Watson-Watt's earlier research work had been concerned with cathode ray direction finders for locating thunderstorm flashes, 252,263 (1926), see also Canadian 251,024 (1925), McNAUGHTON, ANDREW and STEEL, W. A. He turned to the study of radio location of aircraft when in 1935 he was asked to report on whether a death-ray could reasonably be devised using radio waves. He replied to the effect that while radio-destruction was not likely, radio-detection might be.

249   Radar beam scanning
909,924 (1962)   New or improved method of radio scanning. TUCKER, DAVID GORDON and DAVIES, DAVID EVAN NAUNTON, National Research Development Corporation

250   Radial-ply tyres
628,060 (1949)   Improvements in and relating to pneumatic tyres. MANUFACTURE DE CAOUTCHOUC MICHELIN

This was the first steel cord radial tyre, commercially well known as 'Michelin-X'. The first textile cord radial tyre, the Pirelli 'Cinturato', was covered by 700,435 (1953), PIRELLI SOCIETA PER AZIONI.

251   Radio
For early Marconi patents see under WIRELESS TELEGRAPHY.

252   Radio beacons
594,530 (1947)   Improvements in radio navigational systems. EARP, CHARLES WILLIAM and STRONG, CHARLES ERIC, Standard Telephones and Cables Limited

253   Radio receivers
The 'super regenerative' receiver   182,135 (1923)   Improvements in or relating to wireless signalling systems. ARMSTRONG, EDWIN HOWARD
The 'super heterodyne' receiver   137,271 (1920)   Improved method for receiving wireless signals. ARMSTRONG, EDWIN HOWARD

254   Radio valves and microwave tubes
Diode   24,850/1904   Improvements in instruments for detecting and measuring alternating electric currents. FLEMING, JOHN AMBROSE
Triode   1,427/1908   Improvements in space telegraphy. FOREST, LEE DE. See also 100,959 (1917), FOREST, LEE DE
Pentode   287,958 (1928)   Improvements in or relating to circuit arrangements and discharge tubes for amplifying electric oscillations. Wade, Harold for N.V. PHILLIPS' GLOEILAMPENFABRIEKEN
Variable mu   382,945 (1932)   Improvements in or relating to eletric discharge tubes. BOONTON RESEARCH CORPORATION
Klystron   523,712 (1940)   An improved electrical discharge system and method of operating the same. VARIAN, RUSSELL HARRISON, Leland Stanford Junior University
Cavity magnetron   588,185 (1947) Improvements in high frequency electrical oscillators. RANDALL, JOHN TURTON and BOOT, HENRY ALBERT HOWARD, The University, Edgbaston, and WRIGHT, CHARLES SEYMOUR, Admiralty, London
Travelling wave tube   623,537 (1949) Improvements in or relating to electron discharge devices, KOMPFNER, RUDOLF, Clarendon Laboratory, and WRIGHT, CHARLES SEYMOUR, Admiralty, London

255   Radio-wave lenses and mirrors
860,826 (1961)   Improvements in or relating to electromagnetic wave lenses and mirrors. JONES, SPENCER SELTH DUNIAM; GENT, HUBERT and BROWNE, LANGHOR, Ministry of Aviation

This interconnected aerial system is known as the 'boot-lace' aerial.

256   Rapid freezing
292,457 (1929)   Improvements in or relating to the freezing of food substances. BIRDSEYE, CLARENCE, General Foods Company. See also 311,247 (1929), 336,632 (1930), 336,634 (1930) and 336,637 (1930).
1,029,254 (1966)   Method and apparatus for flash freezing. INTEGRAL PROCESS SYSTEMS, INC.

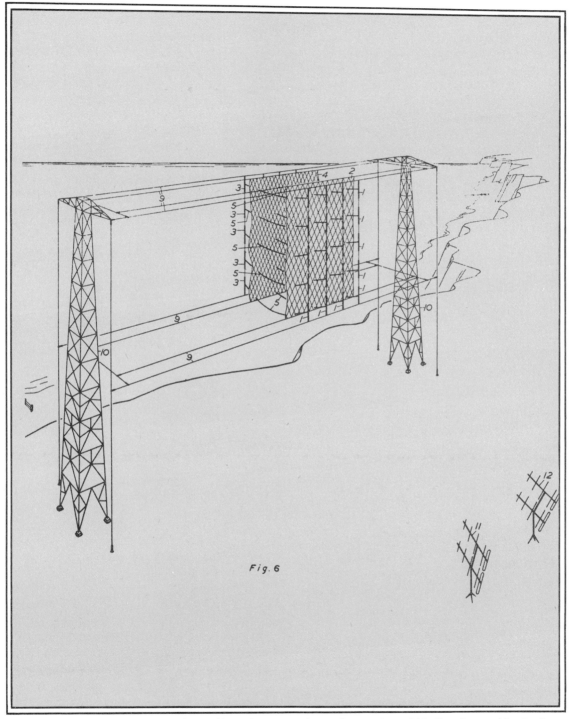

Fig. 6

255 One object of this invention was 'to provide an electromagnetic wave lens or mirror which allows improved freedom in design'. Several aerial configurations are described, the one shown here is for use in a VHF forward-scatter propagation aerial system.

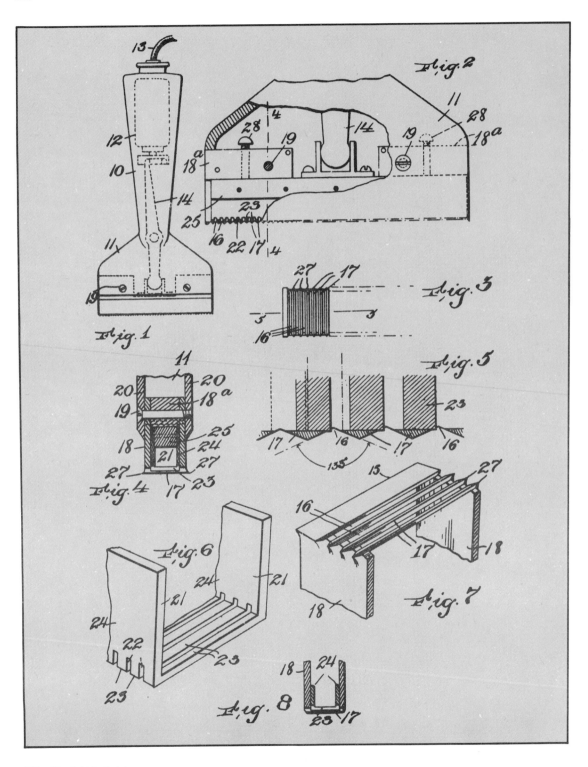

257   The Schick electric razor.

The Birdseye patents led to the rapid development of the frozen food industry after the Second World War. New methods of ultra-rapid freezing have been developed in more recent years; the second patent above describes one such method.

257 Razor, electric
326,374 (1930)   An improved shaving machine. SCHICK, JACOB

258 Reaper
United States 21 June 1834. Improvements in machines for reaping small grain. McCORMICK, CYRUS H. United States patents were not numbered at this time.

McCormick completed his first reaper in 1831 but it was not until 1833 that he could find a farmer to put it to practical use. He took out further patents in 1845 and 1847. In the last of these he included a seat for the operator. The improved machine was shown in the 1851 Exhibition in London where it was awarded a medal as the most valuable article contributed.

259 Recoil-damped guns
125,126 (1919)   Improvements in non-recoiling guns suitable for use on aircraft. HADCOCK, ALBERT GEORGE and FORSTER, GEORGE SIR W. G., Armstrong Whitworth & Co. Ltd

Relates to non-recoiling guns used on aircraft and from which, simultaneously with a projectile, a compensating mass is discharged rearwardly.

260 Regenerative furnace
2,359/1867   Manufacture of cast steel. SIEMENS, CHARLES

Improvements in furnace efficiency was obtained by using the heat of outgoing flue gases to heat incoming air. The regenerator principle was first

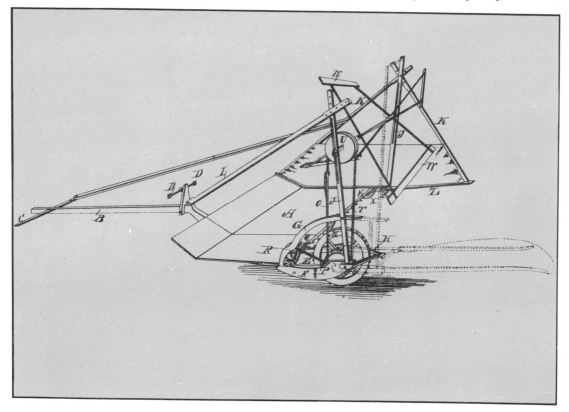

258 The McCormick reaper, surely one of the most significant agricultural inventions of all time.

proposed by ROBERT STIRLING, 4,081 (1816). In his engine patent Stirling had mentioned a possible application for glass furnaces but was aware that it could be applied to other furnaces as is witnessed by the fact that in his original manuscript specification he wrote in 'iron furnaces' and crossed it out. JAMES BEAUMONT NEILSON's hot-blast furnace for smelting had also applied the Stirling principle.

261   Refrigeration

First British patent for cooling foods   4,331 (1819)   Certain improvements and sundry apparatus for cooling and condensing and ventilating worts, liquors and all other fluids or solid matters. SALMON, ROBERT and WARREL, WILLIAM
Early vapour compression cycle patents   6,662 (1834)   Apparatus for producing ice and cooling fluids. PERKINS, JACOB
747/1856   Producing cold by the evaporation of volatile liquids in vacuo, the condensation of their vapours by pressure and the continued re-evaporation and recondensation of the same materials. HARRISON, JAMES. See also 749/1856 and 2,362/1857.

WILLIAM CULLEN, a lecturer in chemistry at the College of Glasgow, in an essay written in 1755, recorded experiments in which freezing temperatures were obtained by the evaporation of nitrous ether under reduced pressure. Perkins's machine was the first successfully to put this principle to practical use. The use of refrigeration in ships was introduced by the Frenchman CHARLES TELLIER, French patent 81,858 (1868). The first cargo of frozen meat was carried between the Argentine in the vessel *Le Frigorifique* designed by Tellier. For later developments see entry on RAPID FREEZING.

262   Reinforced concrete

2,293/1854   Improvements in the construction of fireproof dwellings, warehouses and other buildings, or parts of the same. WILKINSON, WILLIAM PORTLAND

This patent proposes setting wire ropes, bent over or frayed at the ends, into concrete ceilings. This came much closer to today's concept of reinforced concrete than other engineering practices at the time. HENRY FOX, 10,047 (1844) had proposed the use of spaced iron beams to support a concrete structure in which part of the beams were encased in concrete. There were several French engineers using combinations of concrete and metal for floors, see for example 2,659/1855, CIOGNET, FRANÇOISE, but in these there was no suggestion of any interaction between the concrete and the metal. One other interesting French conception was contained in a provisional specification lodged with the Patent Office in 1855 by JOSEPH LOUIS AMBOT for 'an improved building material as a substitute for wood'. This involved the embodiment of wire mesh in concrete. Ambot exhibited a boat at the Paris Exhibition of 1855 in which the shell was constructed of two-inch thick concrete strengthened with such a mesh.

263   Respirator

6,988 (1836)   Improvements in curing or relieving disorders of the lungs. JEFFREYS, JULIUS

264   Revolvers

13,527/1851   Improvements in rifles and other fire-arms. ADAMS, ROBERT

A self-cocking revolver in which the barrel and frame were forged in one piece. This dominated the British industry at the time and there were a number of significant developments after its expiry in 1865. In the United States the main producer at the time was Colts. For an early example of their products see 6,909 (1835), COLT, SAMUEL.

265   Road sweeper

9,433 (1842)   Apparatus for cleaning and repairing roads applicable to other purposes. WHITWORTH, JOSEPH

266   Rolling of metals

1,351 (1783)   Machinery, furnace, and apparatus; for preparing, welding and working various sorts of iron. CORT, HENRY

Rolling, as such, was not unknown before this

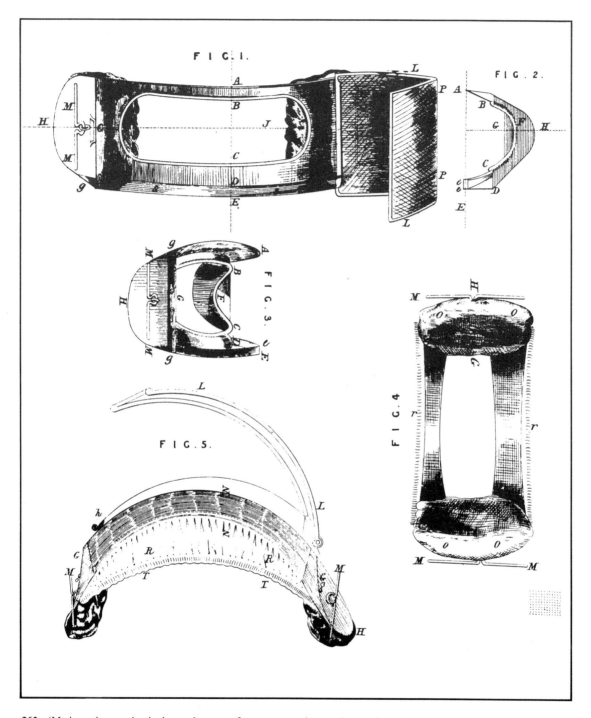

263 'My invention consists in the employment of apparatus to abstract the heat from the breath during the act of expiration (which it receives from the lungs and from the air passages), and give off or transfer such heat to the incoming air which is drawn in during the act of inspiration, and thus warm the air, and render it unirritating to the bronchial and other pulmonary surfaces.... To obtain this object it is well known that invalids are sent to temperate parts of our island and to warm climates.'

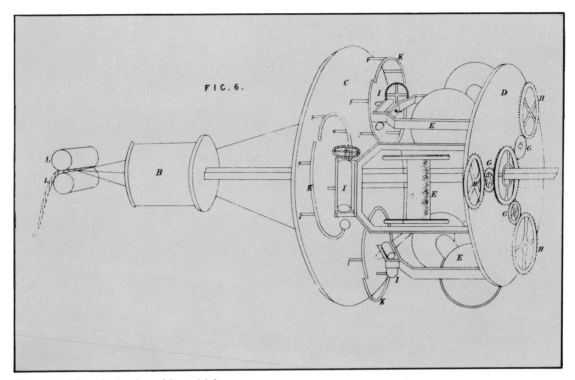

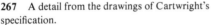

267 A detail from the drawings of Cartwright's specification.

time. Plain rolling had been in use in the tin plate industry and the 'common mill' of the time rolled or slit iron sheets by hammering. In Cort's method the iron was broken down into merchantable shapes by rolls cut with grooves and collars meshing with one another.

267 Rope making

1,876 (1792) Machinery for manufacturing wool, hemp, flax; silk, hair and cotton, into yarn, twist, ropes and cables, and until perfected in the loom, and cut for raising a pile. CARTWRIGHT, EDMUND

To make a strand from a fibre yarn as much twist has to be put into the fibres as is taken out by the stranding. In the cordelier described in this patent this is effected automatically by epicyclic gearing.

268 Rotary crank

1,263 (1780) Machine for boring, turning, rolling, grinding corn, and all sorts of grinding; forging, flatting, and slitting of iron, and other

work that a mill is capable of performing by a rotative motion. PICKARD, JAMES

This patent covers the crank mechanism for converting reciprocating motion into rotary motion. It was followed a year later by a patent by JAMES WATT, 1,306 (1781) which covered various crank equivalents including the sun-and-planet gear.

269 Rotary lawn mowers

Rotary blade with fan to provide suction
336,079 (1930) Improvements in lawn mowers. Villiers, Arthur Henry for BEAZLEY, WILLIAM E. The United States patent is 1,827,559 (1931).
Blades shaped to give downdraft to support mower 929,610 (1963) Improvements in lawn mowers. AKTIEBOLAGET FLYMO, Sweden. See also 992,120 (1965), HANSOM, BERNERD STEPHEN, Wilkinson Sword Ltd

This also describes a mower supported on a cushion of air but an amendment added later disclaims the hovercraft principle accepting the pre-disclosure in 929,610.

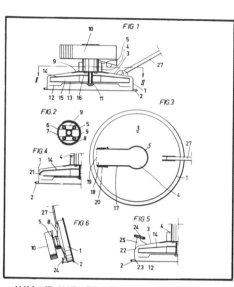

WHAT WE CLAIM IS:—

25   1. A lawn mower comprising a cutter, a motor for driving the cutter, a housing open at one side, and a blower arranged to be driven by said motor, the arrangement being such that when the mower is in 30 operation the open side of the housing is adjacent to the ground and the blower produces an air cushion below the housing to support the mower clear of the ground.

  2. A mower according to claim 1, 35 wherein the cutter is adapted to rotate about a vertical axis when in operation, and the cutter and the blower are mounted on a common shaft.

  3. A mower according to claim 2, 40 wherein the common shaft is an extended shaft of the motor.

  4. A mower according to claim 2 or 3, wherein a friction coupling is provided between the cutter and the shaft.

45   5. A mower according to claim 1, 2, 3, or 4, wherein the housing is substantially cylindrical and the motor has a shaft which is co-axial with the housing and which extends through an air inlet opening at the 50 middle of the side of the housing opposite to the open side.

  6. A mower according to claim 5, wherein the housing is provided with an external strengthening flange around its open side. 55

  7. A mower according to claim 5 or 6, wherein the housing is provided with a flange extending partly around the air inlet opening, leaving a gap for the entry of air.

  8. A mower according to claim 7, 60 wherein a wire mesh or grid extends across the said gap and is held in position by a spring encircling the flange.

  9. A mower according to claim 7, wherein an air inlet channel extends from 65 the gap radially on the housing, and the inlet end of the channel is covered by a wire mesh or grid.

  10. A mower according to any of the preceding claims, wherein a handle for mov- 70 ing and steering the mower is pivotably attached to the housing.

  11. A mower according to any of the preceding claims, comprising a wheel on which the mower can be balanced and trans- 75 ported when not in operation.

  12. A mower according to any of the preceding claims, wherein the blower comprises two axially-spaced circular members interconnected by vanes. 80

  13. A mower according to claims 2 and 12, wherein the air leaves the blower adjacent to the outer parts of the cutter.

  14. A mower according to claim 12 or 13, wherein the axial spacing of the two 85 circular members increases from the periphery towards the axis of the blower.

  15. A mower according to claim 12, 13 or 14, wherein one of said circular members has a central circular opening, and the 90 vanes extend from the edge of this circular opening to the peripheral edges of the circular members.

  16. A lawn mower constructed and arranged substantially as hereinbefore de- 95 scribed with reference to and as illustrated in the accompanying drawing.

HANS & DANIELSSON,
Chartered Patent Agents,
51, King Square,
London, E.C.1.

269   Here we show the drawings for the Flymo lawnmower together with the list of claims. In these claims, which always come at the end of the written part of a specification, the patentee carefully defines the extent of the monopoly that he is claiming. The successful outcome of any legal dispute over the invention may well depend upon a court's interpretation of the wording of these claims.

270   Rubber heels for shoes
2,384/1854   Manufacture of articles of
caoutchouc or form compositions of which
caoutchouc forms a component part. ROSS,
GEORGE

271   Safety fuse
6,159 (1831)   Instrument for igniting gunpowder
when used in blasting rocks, and in mining
('Miner's safety Fuze'). BICKFORD, WILLIAM

272   Safety razor
28,763/1902   Improvements in or relating to
safety and other razors. GILLETTE, KING
Gillette produced the idea of the safety razor with
the disposable blade. Another inventor WILLIAM
NICKERSON, working with Gillette, perfected the
blade.

273   Sanatogen
1,541/1898   Albumen preparations, soluble in
water, from glycerophosphates and albuminoids.
IMRAY, OLIVER

274   Sapphire needle for gramophone
603,606 (1948)   Improvements relating to styli
for sound reproducing. KILLICK, MARIE LOUISE

275   Saw-tooth generator
Miller integrator type   575,250 (1946)
Improvements relating to thermionic amplifying
and generating circuits. WHITELEY, JOSEPH WILLIAM,
A. C. Cossor Ltd
Resonance-return type   400,976 (1933)
Improvements relating to oscillatory electric
circuits, such as may be used, for example, in
connection with cathode ray devices. BLUMLEIN,
ALAN DOWER, Electrical and Musical Industries
Ltd

These are applied for spot deflection in television
and other cathode ray tube apparatus.

276   Screen printing
756/1907   Improvements in and relating to
stencils. SIMON, SAMUEL

This is the first known patent for screen printing,
although experimental work using fabric as a
screen for printing was undertaken in France and
Germany, as far back as 1870. A marked advance
in techniques was made in 1929 when LOUIS F.
D'AUTREMONT introduced the use of shellac film
tissues.

277   Screws
4,117 (1817)   Making screws of iron, brass, steel,
or other metals, for use in woodwork. COLBERT,
JAMES GERARD

This patent was bought up by JOHN SUTTON
NETTLEFOLD and influenced the formation of the
most famous firm of British screw makers.
Colbert's machine was very soon superseded by
machines from the United States which were
exhibited at the 1851 Exhibition.

278   Sealed-beam headlights
479,258 (1938)   Improvements in and relating
to light projecting devices. BRITISH THOMSON-
HOUSTON COMPANY LTD

279   Seed strips
510,136 (1939)   Improvements in and relating to
the storage and sowing of seeds. HARTLEY, VINCENT
See also earlier patent by same inventor 488,392
(1938). Refers to procedure for fixing seeds
between two layers of paper which disintegrate
quickly on exposure to moisture from the soil.

280   Seismic exploring
United States 2,732,906 (1956)   Seismic
surveying. MAYNE, WILLIAM HARRY, Olive S. Petty

281   Self-winding watch
218,487 (1924)   Improvements relating to wrist
watches. HARWOOD, JOHN
Self-winding pocket watches working on the
pedometer principle in which the mainspring was
wound up by a small swinging weight were known
well before this time. Harwood, whose interest in a
self-winding mechanism stemmed mainly from a
desire to seal the watch from dirt which otherwise
might enter through the winding mechanism, used
a small weight swinging through an arc. This
patent was forerunner of presentday developments
in self-winding watches.

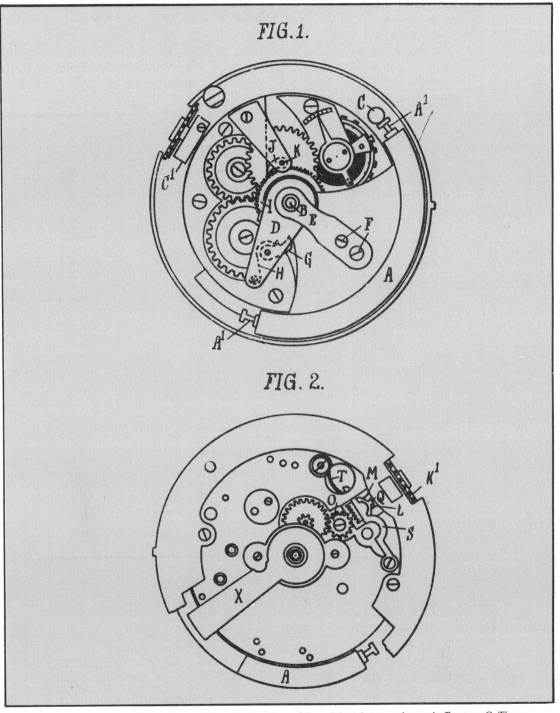

281 The power for winding the watch comes from the vibrations of the weight A between the two buffer stops C. These vibrations are activated by the movements of the wearer and wind up the main spring of the watch by means of a pawl acting on a ratchet wheel.

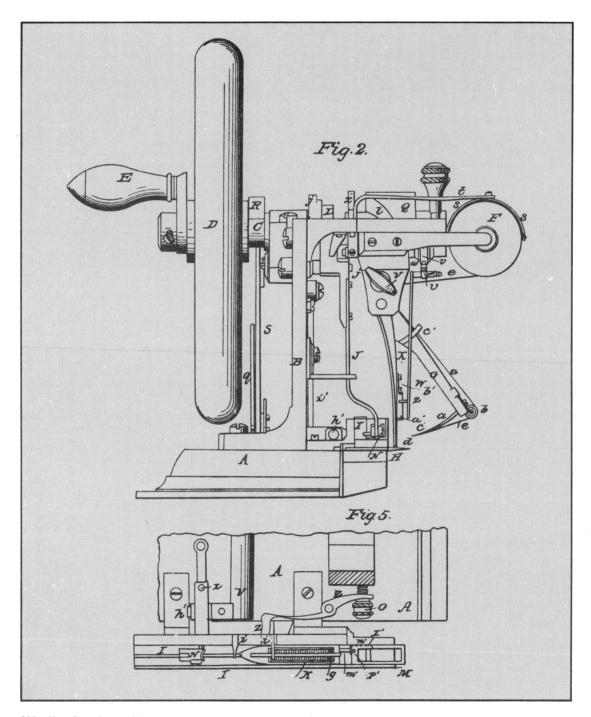

282   Howe's sewing machine.

282   Sewing machines

1,764 (1790)   An entire new method of making and completing shoes, boots, splatterdashes, clogs and other articles by means of tools and machines also invented by me for that purpose, and of certain compositions of the nature of Japan or varnish which will be very advantageous in many useful applications. SAINT, THOMAS
United States 4,750 (1846)   Improvements in sewing machines. HOWE, ELIAS

Saint's patent was the first for a working sewing machine. It worked on the chain stitch principle and there is some doubt that more than one such machine was ever constructed. The next notable development was in France where BARTHELEMY THIMONNIER obtained a patent in 1830 for a model which was sufficiently good to be used to equip a factory making uniforms for the French Army. He came to England in 1848 and obtained English patent 12,060 (1848), registered in the name of his then partner, JEAN MARIE MAGIN. Thimonnier's machines also worked on the chain stitch principle. The first invention of the lock stitch method was due to WALTER HUNT in the United States in about 1833. Hunt's early machine was never patented and appears never to have reached the stage of working practically. Howe's patent above described the first successful lock stitch machine. Like Thimonnier in France he had great difficulty selling his machine, meeting with considerable resistance from the trade, sometimes violent in nature. ISAAC MERRITT SINGER, on the other hand, who took out his first patent, United States 8,294, in 1851, was able to commercialise his improved machine much more effectively. The principle of the rotary hook and reciprocating bobbin was introduced by ALLEN B. WILSON, United States 8,296 (1851). The first patents for an electrically operated machine were United States 118,537–8 (1871), JONES, SOLOMON, but the first successful electrical machines were made by Singer in 1889.

283   Sextant

550 (1734)   Quadrant. HADLEY, JOHN

284   Silicones

542,655 (1942)   Methyl silicone condensation products. ROCHOW, E. G., British Thomson-Houston Company Ltd

Much of the early work on the study of silicones was undertaken in the first half of this century by Professor F. S. KIPPING of Nottingham University. The first commercial application was in glass-fibre insulation. The Rochow direct process of producing silicones is now used for producing most of the commercial silicones, being safer, and more economic, than the Grignard reaction process previously used.

285   Silk spinning

422 (1718)   Three engines, one to wind the finest raw silk, another to spin, and another to twist the finest Italian raw silk into organzine. LOMBE, THOMAS

Sir Thomas Lombe erected a factory in Derby in 1718. His machines owed much to others, used for doubling and twisting, that were already in use in Italy.

286   Silver plating

8,447 (1840)   Improvements in coating, covering or plating certain metals. ELKINGTON, GEORGE RICHARD and ELKINGTON, HENRY

287   Single sleeve valve

18,140/1910   An improved internal combustion engine, BURT, PETER, Argyll's Limited

Single sleeves were ousted by poppet valves for car and motor cycle engines, but were successfully adapted by the Bristol Aeroplane Company for radial aero-engines mass-produced during the Second World War, see 293,409 (1928), FEDDEN, A. H. R. and BUTLER, L. F. G., Bristol Aeroplane Company.

288   Snopake

837,813 (1960)   Method of correcting copy material. BATTELLE DEVELOPMENT CORPORATION

289   Soda production

3,131/1863   Apparatus by means of which the formation of carbonates of soda by direct combination is rendered practically available for manufacturing purposes. SOLVAY, ERNEST

See also improvements to the process described in 1,525/1872. Before this time soda was produced, in the main, using the Le Blanc process, covered by French patent 9 (1791), LE BLANC, NICOLAS.

290   Soda water

3,232 (1809)   A new mode of preparing soda and other mineral waters, spiritous, acetous, sacharine, and aromatic liquors, and sundry improvements relating thereto. HAMILTON, WILLIAM FRANCIS

Soda water was first manufactured in Dublin by the family firm of A. and R. Thwaites in 1799 and first introduced to the public by ROBERT PERCIVAL, a professor of chemistry at Trinity College, Dublin, in 1800. Hamilton's contribution was to develop a continuous process of manufacture. He worked for JOSEPH BRAMAH who may well have contributed to the design of the machine.

291   Spinning

562 (1738)   Machine for spinning wool and cotton. PAUL, LEWIS
962 (1770)   Machinery for spinning, drawing and twisting cotton. HARGREAVES, JAMES
931 (1769)   Machinery for the making of weft or yarn from cotton, flax and wool. ARKWRIGHT, RICHARD

Paul's patent was the first to propose the principle of roller drawing. He described an arrangement using two sets of rollers; but there is some doubt whether he was able to incorporate this concept in a practical machine. Arkwright's patent proposed four sets of rollers and there is no doubt that he was able to solve the practical difficulties of roller drawing and also combine many other concepts to produce practical machines. The 'jenny' of Hargreaves was essentially a direct attempt to make it possible for one operative to control a number of spinners without basically changing the working principles of the spinners themselves.

292   Spinning box

23,157/1900   Improvements in apparatus for use in the production of textile fibres, or filaments, from solutions of cellulose, or of other material from which fibres or filaments can be formed. TOPHAM, CHARLES FRED
23,158/1900   Improvements in means to be employed in twisting fibres, or filaments, and putting them into coil form. TOPHAM, CHARLES FRED

The second of these two related patents deals with the well-known Topham spinning box, which uses centrifugal force to twist and coil artificial or natural fibres into yarn.

293   Sponge rubber

1,111/1914   A new or improved rubber substance and process for making the same. SCHIDROWITZ, PHILIP and GOLDSBROUGH, HAROLD

294   Spray drying

United States 125,406 (1872)   Process of drying and concentrating liquid substances by atomising. PERCY, S. R.

This is the first known patent.

295   Stainless steel

United States 1,197,256 (1916)   Cutlery. BREARLEY, HARRY

Harry Brearley told his own story of the invention of stainless steel in an article in *The Sheffield Daily Independent* of 2 February 1924. At the time of the discovery Brearley was in charge of a research department of Thomas Firth and Sons Ltd. The first steels with the right properties were the results of attempts to find a suitable steel for use as inner linings for large gun barrels. Brearley saw that similar steels could be used for making stainless cutlery but arguments arose between him and his employers as to the suitability of such steels for this purpose. These arguments became sufficiently bitter to cause Brearley to resign his employment

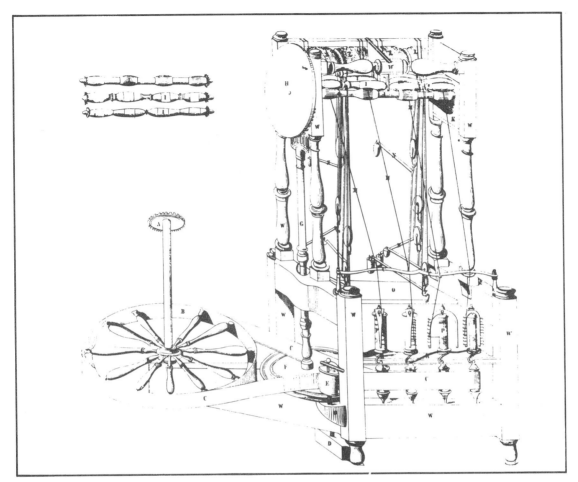

291   The well-known spinning jenny.

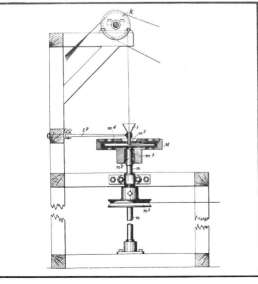

292   Topham's spinning box. Here the centrifugal force
imparted on the thread by the spinning of the box M gives the
thread the twist that is so essential if it is to be satisfactorily
spun into yarn.

and seek work elsewhere. It was mainly due to this difference of opinion that no British patent was applied for. The United States patent was taken out on Brearley's behalf by an American friend JOHN MADDOCKS, and the rights to it were eventually sold by them to FIRTH AND SONS who set up the FIRTH-BREARLEY STAINLESS STEEL SYNDICATE. Brearley continued to complain that Firth and Sons were claiming too great a share of the credit for the invention.

296   Starting blocks
United States Patent 1,709,832 (1929)   Starting block. BARRON, ALBERT M.

297   Steam engines
356 (1698)   Raising water and giving motion to mill works by the impellant force of fire, useful for draining mines, serving towns with water, and working all kinds of mills in cases where there is neither water nor constant wind. SAVERY, THOMAS
913 (1769)   Method of lessening the consumption of steam and fuel in fire engines. WATT, JAMES

There is no specification with Savery's patent but he published an account of his engine privately in 1702 under the title *The Miner's Freind; or, an engine to raise water by fire, described, and of the manner of fixing it in mines, with an account of the several other uses it is applicable to; and an answer to the objections made against it*. He exhibited a model of his engine to the King at Hampton Court and later to the Royal Society in 1699. An account is given in *Philosophical Transactions of the Royal Society*, London, vol. 21, p. 228. The principle of the engine was simple, a vessel, called a receiver, was fixed at a height of not more than 22 to 26 feet above the level from which water was to be pumped, steam was fed to the receiver until it was full. The steam supply was then cut off and the steam in the receiver condensed by a spray of cold water; the consequent reduction of pressure in the receiver would cause water to be drawn into it through pipes leading downwards. This water would later be forced

upwards through other pipes when steam was once again allowed into the receiver. All the cocks and valves were worked by hand. The machine was found to work adequately when the pumping height was limited but when attempts were made to fix the pumps in mines they almost invariably failed, probably because the engineering experience of the day could not produce pipe fittings or containers of sufficient strength to withstand the high steam pressures that would be involved in forcing water to any marked height above the level of the receiver. THOMAS NEWCOMEN was the next to make significant improvements. Newcomen's engines made no attempt to use the steam to force the water upwards, they made use of atmospheric pressure to produce a vertical up and down motion which could be used for pumping. An open-topped cylinder fitted with a piston was placed directly above the boiler and above this was mounted a beam or lever pivoted centrally and coupled at one end to the cylinder rod and at the other to the pump rods. Steam was admitted to the cylinder below the piston and then condensed by the injection of cold water. A vacuum was thus created and the piston was forced downwards by atmospheric pressure drawing the pump rods up, the weight of the pump rods themselves were sufficient to lift the piston again so the internal pressures need never exceed that of the atmosphere. Newcomen never applied for a patent but built many successful engines. While Savery was alive there is no evidence that any royalties were paid to him for the Newcomen engines. When Savery died his patent rights were bought up by a consortium of business men who extracted royalty payments on the Newcomen engines. Newcomen engines were extravagant on the use of fuel. This was not a serious defect at the mines where fuel was plentiful and to hand, but made their use in other locations uneconomic. Watt, by adding a separate condenser to the Newcomen engine, made the engine much more efficient. The Watt patent noted above was the first for the separate condenser engine, for the later double-acting engine see 1,321 (1782).

58 THE STEAM ENGINE.

" well as the common, and he thinks it absolutely certain ;" but his cylinder has been so badly bored, that it is one-eighth of an inch wider at one end than the other.

Proposes to work the condenser by a water-wheel (a). Thinks cylinders of block tin will be the cheapest and best; will not be liable to be dissolved by oil; will not wear; pistons need not be more than a quarter of an inch thick. Suggests oil pump to return oil constantly to piston.

Enrolls his Specification, which discloses an improvement in the *method of forming a vacuum under a piston*, that marks an era in the history of mechanical inventions. The document appears to have been drawn up by Small and Boulton.

A.D. 1769, January 5.—No 913.

WATT, JAMES.—Lessening the consumption of steam and fuel in fire engines.

1. Keeping the steam vessel, or cylinder, as hot as the steam that enters it, by casing it with substances that conduct heat slowly, or with heated bodies ; and by preventing substances colder than the steam from entering or touching it.

2. Condensing the steam in vessels [condensers] distinct from the cylinder, but communicating with it ; and keeping the condenser as cold as the air in its neighbourhood by the application of cold bodies.

3. Drawing out with pumps, wrought by the engines, the uncondensed elastic vapour and air that may remain in the cylinder and condenser.

4. Using the expansive force of steam to press on pistons ; and working an engine by this force alone, and discharging the steam into the air after it has done its office.

5. A rotary engine. The steam vessel has the form of a hollow ring mounted on a horizontal axle, with proper inlets and outlets for the steam ; and valves that suffer a weight or fluid metal that fills up a portion of the channel, to go freely round, steam-tight, and in one direction only. On steam being admitted between the weight and a valve, the weight moves in that direction in which the valve opens, and gives a preponderance to that side of the ring

Muirhead, Inventions of James Watt, vol. i., p. 44

THE STEAM ENGINE. 59

up which it is forced, and produces a circular motion ; and, as the wheel moves round, the steam that has raised the weight is drawn into the condenser, or discharged into the air.

6. Partially condensing the steam, and working an engine by its alternate expansion and contraction.

7. Using oil, fat, wax, resinous bodies, and fluid metals to render the piston and other parts air and steam tight.

[Printed. 3d. Repertory of Arts, vol. i. p. 217 ; Mechanics' Magazine, vol. i. p. 4 ; Register of Arts and Sciences, pp. 24, 346 ; Engineer and Mechanics' Encyclopedia, vol. ii. p. 375 ; Webster's Reports, vol. i. p. 56, note, also pp. 230, 232, 255 ; Webster's Patent Law, p. 46, also p. 127, cases 36, 21, 28 ; Blackstone's Reports, vol. iii. p. 463 ; Carpmael's Reports on Patent Cases, vol. i. pp. 117, 156 ; Davies on Patents, pp. 155, 168, 221 ; Parliamentary Report, 129 [Patent Law], pp. 187, 189, 196 ; Vesey, jun., Reports, vol. iii. p. 140 ; Term Reports, by Durnford and East, vol. viii. p. 95 ; Rolls Chapel Reports, 6th Report, p. 160. Rolls Chapel.]

At this time an engine, with an eighteen-inch cylinder and five-feet stroke, was in process of erection at Kinniel House, where Dr. Roebuck's collieries were situated. This seems to have been considered a testing, final performance. As the machine approached completion, Watt, in a letter to a friend, says, " he is one whose anxiety for his approaching doom keeps him " from his night's sleep, and whose fears at least equal his " hopes." However, in September the Kinniel engine was set to work, and its performance realized Watt's expectations.

Eighteen fire engines on Newcomen's model at work throughout Cornwall exert a total effect of 481 horses power only ; the principal makers being Jonathan Hornblower and Thomas Nuncarrow. At Newcastle, at this same time, there were engines exerting a total of 1,183 horses power.

Watt mentions in a letter, May 28th, a method of " doubling " the effect of steam by using the elastic force of vapour rushing " into a vacuum at present lost." Opens a valve until a fourth only of the cylinder is filled.

A.D. 1770, January 5.—No 949.

CLARKE, DUGALD.

1. Describes the construction of a sugar cane mill, to be " worked by animals, wind, water, fire, or the waves or surge of " the sea."

2. Notices an iron or copper boiler of a fire engine whose lower portion is formed like a hollow cone.

---

26 THE STEAM ENGINE.

A.D. 1698.

WHILE Papin was engaged in his fire engine projects, Mr. Thomas Savery had perfected an engine, in which steam was used to raise water in a different manner from that followed in the Cassel machine ; for this he received a patent.

" A.D. 1698, July 25.—No 356.

SAVERY, THOMAS.—" A new invention for raiseing of water and " occasioning motion to all sorts of mill-work by the impellent " force of fire ; which will be of great use and advantage for " drayning mines, serveing townes with water, and for the working " of all sorts of mills where they have not the benefit of water " nor constant windes."

[No Specification enrolled. Patent printed, price 4d. Register of Arts and Sciences, vol. 3, p. 252. Engineer and Mechanics' Encyclopedia, vol. 2, p. 697. Stuart's History of the Steam Engine, p. 34. Stuart's Anecdotes of Steam Engines, vol. 1, p. 102. Woodcroft on Steam Navigation, p. 1, 1848.]

A.D. 1698.

PAPIN, being desired by the Prince of Hesse " to construct an " engine to raise water by fire in a more advantageous manner " than he had done before," made experiments to accomplish the wish of the Elector. In his renewed attempts he operated wholly by the pressure of the steam. He seems to have failed ; and in November his engine was ruined by the drifting of the ice in the river (Fulda ?), in or near which it was placed. He had wholly abandoned the method of steam tubes and pistons.

In a letter to Leibnitz, of this date, Papin says he had made a model of a *carriage* propelled by this force, which succeeded, and hoped it would answer for *boats* (a).

Papin was present at a trial of Savery's row-boats on the Thames.

A.D. 1699.

The " moulin a feu," described by Guillaume Amontons, is formed of two concentric ranges of cells fixed on an axle. The outer range consists of twelve compartments, which are closed and without communication with other. The inner range is divided

(a) Prosser MSS., Woodcroft's MS. Collection, Great Seal Patent Office. Abridgments of Specifications, Marine Propulsion, p. 303, 1858.

THE STEAM ENGINE. 27

into an equal number of cells ; each communicating with the contiguous cell through a valve opening upwards. Each compartment of the outer range communicates by a small pipe with one of the cells of the inner range. The vertical wheel, formed by the two ranges of cells, has a portion of its periphery exposed to the action of the fire ; and another portion immersed in a cistern of cold water. A fire in the furnace heats the air in the cell exposed to it, which expands, and flowing through the small pipe into the cell it communicates with, in the inner range, forces the water in the inner cell through the valve into the cell above it, which, giving a preponderance to that side of the wheel, it descends and brings another outward cell over the furnace ; the air it encloses is heated in its turn, and raises the water from the inner cell into that next above it, which continues the preponderating movement ; each heated cell passing in its revolution through the cold water in the cistern, the air it contains is refrigerated and condensed, and its sides cooled before they again come opposite the furnace (a).

A.D. 1699.

THE small model that SAVERY set to work before the Royal Society in June 14th of this year, and of which they gave two draughts in their Transactions, is probably more perfect than his first model, on which he applied for a patent (b). It is most likely, also, that shown before the Parliamentary Committee on a bill then before the House, for a prolongation of the term of the patent.

" 10 and 11 Gul. III. 1699.

" AN ACT for the encouragement of a new invention by THOMAS " SAVERY, for raising water and occasioning motion to all sorts of " millwork by the impellent force of fire." Savery, says the preamble, having by much care, pains, and expense, since the grant of letters patent, greatly improved his invention, which is likely to be of great use to the public, but which probably will require many years' time and much greater expense to bring it to

(a) Mémoires de l'Académie des Sciences, p. 208, 1699.
(b) Vol. xxi. p. 228, 1699.

297 These four pages from the early Woodcroft patent abridgment volumes illustrate two crucial episodes in the development of the steam engine.

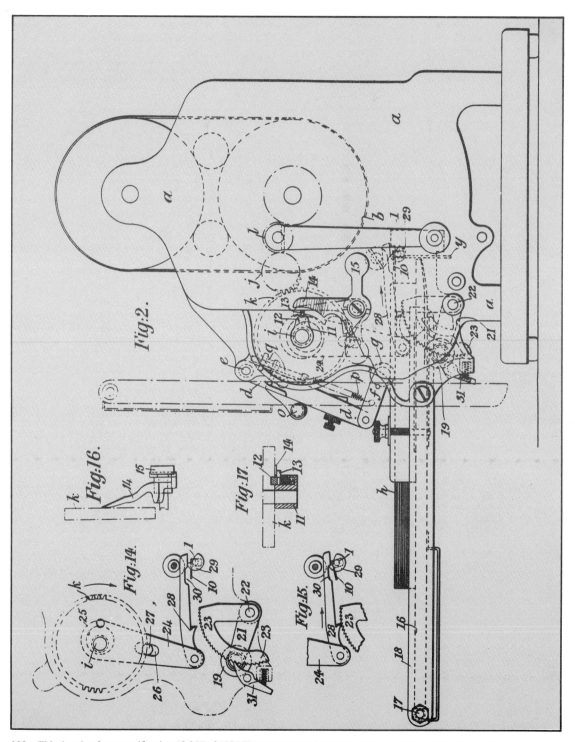

302 This drawing from specification 18,257 of 1905 illustrates an early model of the line of stencil duplicators that enabled document copying to be done in the office as well as the printing works.

298   Steam locomotives
2,599 (1802)   Construction of steam engines; application thereof for driving carriages and for other purposes. TREVITHICK, RICHARD and VIVIAN, ANDREW

Trevithick ran his first locomotive on a prepared track at Penydarran in Cornwall in 1804. By 1808 he had a high-pressure engine running round a circular track laid on a site close to the present Euston station. GEORGE STEPHENSON produced his first locomotive at Killingworth Colliery in 1814. It was Stephenson's locomotives which were later used on the Liverpool and Manchester Railway, the first operational railway to employ locomotive power exclusively.

299   Steel manufacture
Bessemer converter   44/1856   Manufacture of iron and steel. BESSEMER, HENRY
Improvements to cope with ores containing phosphorus   4,422/1877   Improvements in the manufacture of steel, and in the lining of Bessemer converters. THOMAS, SYDNEY GILCHRIST

300   Steel manufacture using coke
380 (1707)   Casting iron bellied pots, and other bellied ware. DARBY, WILLIAM

It was in the manufacture of these pots, first produced at Coalbrookdale, some miles south of Shrewsbury in Shropshire, that Darby perfected the process of smelting using coke instead of charcoal.

301   Steering (radial-axial)
4,212 (1818)   Improvements in axletrees applicable to four-wheel carriages. ACKERMANN, RUDOLF

302   Stencil duplicating
United States 584, 787 (1897)   Duplicating or stencil printing machine. LOWE, H. W.
23,406/1900   Twin cylinder rotary stencil duplication machine. GESTETNER, DAVID. See also 25,373/1901; 14,303/1902 and 18,257/1905.

These were the key patents for the rotary duplicating machines which were to bring stencil duplicating right into the modern office. The rotary duplicating process had been established before by H. B. THOMPSON, 12,011/1889, but on a web basis for the purpose of duplicating patterns for wallpaper. Gestetner had been in the field long before, having been the first to develop an efficient method of duplicating handwritten text. He introduced the cyclostyle pen, 2,450/1881, extending the principle so that it could be used for musical manuscripts, 7,536/1885, and was also the first to propose the use of Japanese waxed paper for preparing stencils, United States 332,890 (1885).

303   Stereoscopic camera
2,064/1856   Improvements in photographic cameras, and in apparatus connected therewith. DANCER, JOHN BENJAMIN

This camera had added distinctions, it was one of the first magazine loading types, it was the first to be fitted with a spirit level and it was also fitted with a special designed double rotating shutter. SIR CHARLES WHEATSTONE, SIR DAVID BREWSTER and others made suggestions regarding the correct distance by which the two lenses should be separated. Dancer himself decided that the distance between a normal pair of human eyes was right. Dancer has further claims to fame. In 1960 he was posthumously awarded a medal by the American Microfilm Association as the inventor of microphotography. He first demonstrated this technique in 1839 but never patented the idea. One of his many interesting works was a series of microphotographs of Queen Victoria's family which were mounted together in a ring, the stone of which acted as a magnifying glass. He was famous as an instrument maker and, among many other noteworthy items, made two of the thermometers JAMES PRESCOTT JOULE used to determine the mechanical equivalent of heat and JOHN DALTON'S last microscope.

304   Stereotype
1,431 (1784)   A method of making plates for the purpose of printing by or with plates instead of the movable types commonly used, and for the vending and disposing of the said printing plates, and the books or other publications therewith

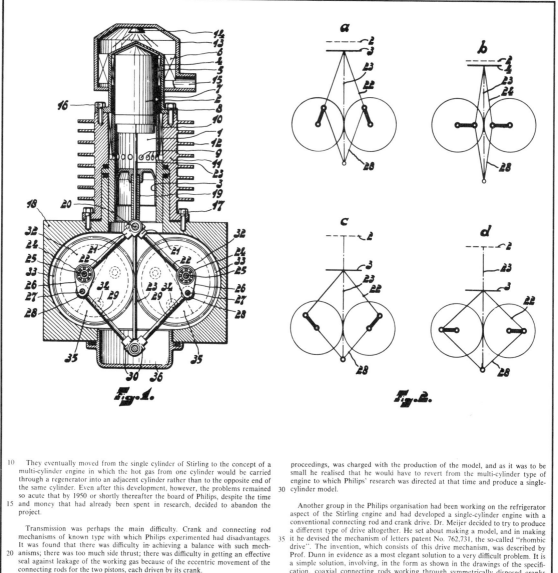

Fig. 1.

Fig. 2.

10   They eventually moved from the single cylinder of Stirling to the concept of a multi-cylinder engine in which the hot gas from one cylinder would be carried through a regenerator into an adjacent cylinder rather than to the opposite end of the same cylinder. Even after this development, however, the problems remained so acute that by 1950 or shortly thereafter the board of Philips, despite the time
15   and money that had already been spent in research, decided to abandon the project.

Transmission was perhaps the main difficulty. Crank and connecting rod mechanisms of known type with which Philips experimented had disadvantages. It was found that there was difficulty in achieving a balance with such mech-
20   anisms; there was too much side thrust; there was difficulty in getting an effective seal against leakage of the working gas because of the eccentric movement of the connecting rods for the two pistons, each driven by its crank.

In November, 1953, a Mr. Rinia, who had been a pioneer in the development of the Stirling engine by Philips, was to celebrate his twenty-fifth anniversary with

proceedings, was charged with the production of the model, and as it was to be small he realised that he would have to revert from the multi-cylinder type of engine to which Philips' research was directed at that time and produce a single-
30   cylinder model.

Another group in the Philips organisation had been working on the refrigerator aspect of the Stirling engine and had developed a single-cylinder engine with a conventional connecting rod and crank drive. Dr. Meijer decided to try to produce a different type of drive altogether. He set about making a model, and in making
35   it he devised the mechanism of letters patent No. 762,731, the so-called "rhombic drive". The invention, which consists of this drive mechanism, was described by Prof. Dunn in evidence as a most elegant solution to a very difficult problem. It is a simple solution, involving, in the form as shown in the drawings of the specifi-cation, coaxial connecting rods working through symmetrically disposed cranks
40   of equal length, themselves connected to two discs rotating on shafts in syn-chronism and in phase but in opposite sense. At one blow this development pro-vided the solution to all the drive problems which had been harrassing Philips.

As a result of this development the board decided to reverse their decision and to continue research into the engine.

306   The few paragraphs above, which describe one of the more intriguing episodes in the history of the Stirling engine, have been copied from the report of a patent case seeking to extend the life in England of Philips' patent 762,731. The drawings above have been copied from the specification. No extension was granted because, although it was accepted that the invention was of considerable merit and that there had not been sufficient time to develop the patent to the point of obtaining any significant commercial return [both these conditions are necessary for the grant of an extension], there had been no effort to encourage the development of the engine in the United Kingdom, i.e. to 'work' the patent in this country. This extract has been copied from *Reports of Patent, Design, Trade Mark and other Cases* which is published by the Patent Office and contains accounts of selected cases chosen by a legal editor.

printed, whereby a much greater degree of accuracy, correctness, and elegance will be introduced in the publication of the works both of the ancient and modern authors than had hitherto been attained. FOULIS, ANDREW and TILLOCK, ALEXANDER

305 Stereophonic gramophone

394,325 (1933) Improvements in and relating to sound transmission, sound recording and sound-reproducing systems. BLUMLEIN, ALAN DOWER, Electrical and Musical Industries, Ltd
This was the first patent to disclose a stereophonic system using a single track, two-channel gramophone recording.

306 Stirling engine

4,081 (1816) Diminishing consumption of fuel; engine capable of being applied to the moving of machinery. STIRLING, ROBERT
605,922 (1948) Improvements in or relating to hot gas reciprocating engines. N. V. PHILIPS' GLOEILAMPENFABRIEKEN
762,731 (1956) Improvements in or relating to hot-gas reciprocating engines and heat pumps and refrigerators operating according to the reversed hot gas engine principle.
N. V. PHILIPS' GLOEILAMPENFABRIEKEN

The Stirling English patent was never printed by the Patent Office or enrolled, although it was duly signed, attested and stamped. The reason for this is not clear. However, it was later published in *The Engineer* of 14 December 1917. As well as describing the basic principles of the engine it explained, in some detail, the 'regenerative' principle which was later put to such good use in NEILSON'S hot-blast furnace and the SIEMENS regenerative furnace. The engine itself, an 'external combustion' engine, was never put to wide use at the time since it was unable to compete with the early steam engines. Philips took up the idea afresh in the late 1930s in an attempt to build power generators which could work on a variety of fuels. The second of the two patents is the first of the Philips patents filed in this country relating to this engine. The third of the patents above describes the 'rhombic drive' mechanism.

307 Stocking frame

First power-driven fully-fashioned frame 7,545 (1838) Machinery for framework knitting.
BARTON, LUKE
Flat frame driven by rotary mechanism 70/1860 Improvements in machinery or apparatus employed in the manufacture of looped fabrics.
COTTON, WILLIAM.
See also 11,255 (1846) and 1,901/1863
Fly-needle-frame 375,371 (1932)
Improvements in the manufacture of fabrics of various kinds by needle action. MORTON, JAMES

The first loom for knitting hosiery was invented by WILLIAM LEE of Calverton near Nottingham about 1590. Queen Elizabeth refused a patent for it. The next significant developments were the invention of the STOCKING RIB and the latch needle, the bearded needle had been incorporated in LEE's early looms. The Cotton Patent machine is the basis of many more modern machines. Warp-knitting machines gradually displaced other machines for many applications after the introduction of the fly-needle-frame.

308 Stocking rib

722 (1758) and 734 (1759) Machine furnished with a set of turning needles, and to be fixed to a stocking frame for making turned rib stockings, pieces and other goods usually manufactured on stocking frames. STRUTT, JEDEDIAH

309 Streptomycin

607,186 (1948) Antibiotic substances and processes for preparing same. MERK AND CO. INC.

This patent describes a new technique for obtaining certain Streptomycin compounds and other antibiotic substances. For an earlier method of producing streptomycin see SCHATZ, A.; BUGIE, E. and WAKSMAN, SELMAN A., *Proceedings of the Society of Experimental Biology and Medicine*, vol. 55, 1944, 66–9. Waksman spent many years studying soil organisms from the point of view of their role in soil processes. In 1939 he turned to a study of their medicinal properties after the discovery of gramicidin by one of his former pupils, RENE J. DUBOIS. He claims to have studied

Forster's Improvements in Gloves for Use in Surgical Operations.

SPECIFICATION in pursuance of the conditions of the Letters Patent filed by
the said Thomas Forster in the Great Seal Patent Office on the 17th October
1878.

THOMAS FORSTER, of the India-rubber Works, Streatham, in the County of
Surrey. "IMPROVEMENTS IN THE MANUFACTURE OF GLOVES OR COVERINGS FOR THE 5
HANDS FOR USE IN SURGICAL OR OTHER OPERATIONS WHERE IT IS ESSENTIAL TO
COVER THE HANDS YET RETAIN DELICACY OF TOUCH."

This Invention has for its object improvements in the manufacture of gloves or
coverings for the hands for use in surgical or other operations where it is essential
to cover the hands yet retain delicacy of touch.   10

I take moulds of plaster, glass, or other suitable material, and after making a
solution of india-rubber of a proper consistence, I carefully dip or immerse, or paint
over the moulds, and then permit them to dry. I repeat the operation until the
covering is sufficiently thick; I place them in a stove and carefully dry them, and
then cure or vulcanize them. The perfected article is then withdrawn from the 15
mould.

Having thus stated the nature of my Invention I will proceed to describe more
in detail the manner in which I prefer to operate.

I make moulds of plaster of Paris (I find this material suits my purpose) by
taking casts from a glove last. Care must be taken to stop the pores of the plaster, 20
and this may be very conveniently done by immersion in gelatine and water,
allowing the plaster in the first instance to absorb as much as it will, and then to
dry in a warm, airy room. The dipping and drying should be repeated until the
mould has a polished surface all over. I then take a solution of india-rubber, about
1 lb. of rubber to 7 or 8 lbs. of solvent, by preference good pure coal oil. I proceed 25
to dip the mould or plaster hand into it; and after the first dipping has dried I
repeat the operation until the desired thickness is obtained. The thinner these
gloves are made the better, so long as they are sufficiently strong to bear handling,
as the object is to cover the hand of the user with an impervious film without
interfering with the delicacy of the touch. The glove is then allowed to dry 30
perfectly in a warm chamber free from dust, and whilst still upon the mould it is to
be vulcanized, for which purpose I dip it into a curing solution consisting of chloride
of sulphur and bisulphide of carbon, by preference, one part of chloride of sulphur
to sixty parts of bisulphide of carbon are suitable proportions.

When dry the glove may be loosened at the wrist, and by turning it over and 35
rolling it down the mould the glove may be stripped off readily.

If by one immersion in the curing solution the glove is insufficiently acted upon
it should be dipped a second time.

The gloves may be cured by exposure to the vapour of chloride of sulphur, but
not so effectually.   40

What I claim is the novel manufacture of impervious vulcanized india-rubber
film gloves, substantially as described.

In witness whereof, I, the said Thomas Forster, have hereunto set my hand
and seal, this Twenty-sixth day of September, in the year of our Lord One
thousand eight hundred and seventy-eight.   45

THOMAS FORSTER. (L.S.)

LONDON: Printed by GEORGE EDWARD EYRE and WILLIAM SPOTTISWOODE,
Printers to the Queen's most Excellent Majesty.
For Her Majesty's Stationery Office.
1878.

312 Surgical gloves. Here we have a patent specification all on one page. It would be difficult to find a modern specification as short as this one. A typical specification would run from three to five pages, although many run considerably longer than this.

some 10,000 cultures before discovering the medicinal properties of streptomycin which had the advantage over earlier antibiotics in that it would attack the gram-negative bacteria against which the penicillin antibiotics were ineffectual.

310   Sulphur, recovery from alkali waste
2,277/1862   Extracting sulphurous acid from oxysulphuret of calcium. Schnell, William for MOND, LUDWIG

311   Sulzer loom
353, 764–5 (1931)   Improvements in looms with nipping shuttles. ROSSMANN, RUDOLF, Teeag Textil Finanz A.G.

This machine differs from the traditional loom in that its shuttle, instead of carrying its own thread, picks up thread at the beginning and drops it at the end of its run across the loom. Several small shuttles are used, each passing across the loom in succession. This early patent did not meet with immediate success; several years of work by Sulzer Bros., in Switzerland were needed before the loom was put into active use in 1950.

312   Surgical gloves
1,532/1878   Improvements in the manufacture of gloves or coverings for the hands for use in surgical or other operations where it is essential to cover the hands and yet retain delicacy of touch. FORSTER, THOMAS

This describes the first successful rubber gloves.

313   Switch, electric
3,256 (1884)   Improvements in or applicable to switches or circuit closers for electrical conducting apparatus. HOLMES, J. H.

This patent covered the loose-handle quick break switch eliminated the dangers of arc-ing across a partially broken contact switch.

314   Switching circuits
Eccles-Jordan circuit   148,582 (1920)   Improvements in ionic relays. ECCLES, WILLIAM HENRY and JORDAN, FRANK WILFRED
Zener diode circuit   752,412 (1956)   Electrical circuits including semi-conductor devices. WESTERN ELECTRIC COMPANY, INC.
For the complementary transistor circuit see 753,014 (1956).
Transistor NOR logic gate   845,371 (1960)   Improved semi-conductor logic units and network composed thereof. BURROUGHS CORPORATION

315   Synchromesh gears
350,334 (1931)   Improvements in or relating to clutches for variable speed gearing. Triggs, William Warren for GENERAL MOTORS CORPORATION

316   Tarmacadam
7,796/1902   Improvements in the means for and the method of 'tarring' broken slag, macadam, and similar materials. HOOLEY, EDGAR PURNELL

317   Teabags
335,157 (1930)   A machine for continuous weaving and filling of pocket-shaped textile fabrics. KOROSY, PAUL VON

318   Teaching machines
1,223,642 (1971)   Improvements in or relating to teaching systems. BITZER DONALD L., University of Illinois Foundation

This is an example of a multi-student teaching machine known as 'PLATO', one of the many sophisticated modern applications of computers.

319   Tear-open tins
1,213,616 (1970)   Improvements in or relating to opening means for containers. COOKSON, WILLIAM, Cookson Sheet Metal

For other early patents in this field see 1,217,295 (1970) and 1,219,084 (1971).

320   Teflon
625,348 (1949)   Tetrafluoroethylene polymers. PLUNKET, ROY J., Kinetic Chemicals Inc.

Plunket found out how to polymerise tetrafluoroethylene by accident, earlier deliberate attempts to do this had always met with failure. He had been assigned by DU PONT to make a study of any possible uses of tetrafluoroethylene as a refrigerant. He synthesised a quantity of the compound and stored it in a cylinder. A few weeks later when he opened up the cylinder some of it had polymerised. Although its unusual properties of inertness were recognised at once, no immediate use for it was found until three years later; in 1944 it was put on the secret list and used to make containers for particularly corrosive materials which were being used in the United States nuclear research programme. Later its uses proliferated. Its non-stick properties were first applied in the bread making industry and applications to domestic kitchenware followed later, see for instance 833,899 (1960), UNION CHIMIQUE BELGE SOCIÉTÉ ANONYME.

321   Telegraph (electric)

7,390 (1837)   Improvements in giving signals and sounding alarms in distant places by means of electric currents transmitted through metallic circuits. COOKE, WILLIAM FOTHERGILL and WHEATSTONE, CHARLES

United States 1,647 (1840)   Improvements in the mode of communicating information by signals by the application of electromagnetism. MORSE, SAMUEL FINLAY BREEZE

JOSEPH HENRY published a detailed description of an electro-magnetic telegraph in 1831 in the United States and similar telegraphs were devised in England, Russia and Germany all at roughly the same time. However, these had to be developed into commercially tenable systems and the necessary developments occurred independently and again almost simultaneously in England and the United States. Successful demonstrations of their different systems were given by Wheatstone and Cooke in England and by Morse in the United States in 1837. In the Wheatstone system letters on a board were indicated by the deflection of five needles, and a calling device was incorporated to draw attention to the observer. They were granted a patent in the United States 10 days before Morse received his but Morse was given priority as the first inventor. Wheatstone and Cooke, however, had the priority in the United Kingdom but Morse

received recompense from many other European governments. The Morse patent describes a prototype of his famous dot-dash code which is still used today.

322   Telephone

4,765/1876   Improvements in electric telephony (transmitting or causing sounds for telegraphing messages) and telephonic apparatus. Brown, William Morgan for BELL, ALEXANDER GRAHAM. The United States patent was 174,465 (1876).

Bell and ELISHA GRAY had been developing a practical harmonic telegraph independently, and both realised that this might lead to the possibility of transmitting speech by wire at about the same time. On St Valentine's day of 1876, Bell filed his patent specification with the United States Patent Office and Gray filed a caveat making very similar claims. In subsequent litigation Bell was given priority, in the United States his patent was interpreted in the broadest sense to cover the basic principle of the transmission of speech by telegraphy, but its interpretation was not so broad in other countries. Bell's first telephones used an electromagnetic device both as a transmitter and a receiver. As a receiver it was reasonably efficient but something more efficient was wanted as a transmitter. THOMAS ALVA EDISON was the first to devise a reasonably efficient transmitter, the flat disk lamp-black microphone, see 2909 (1871), United States 203, 011–9. This was quickly followed by EMILE BERLINER'S patent for a loose contact microphone, United States 463,569 (1891) a caveat for this had been filed before the Edison specification. Edison's patents were taken up by the Western Union Telegraph Company, who had earlier refused an offer of rights to the Bell patent. The BELL TELEPHONE COMPANY acquired Berliner's patents and also employed Berliner to work on improving it. Berliner made good use of the improvements suggested by FRANCIS BLAKE, United states 250, 126–9 (1881). DAVID EDWARD HUGHES had been independently working on loose-contact microphones in England at about the same time.

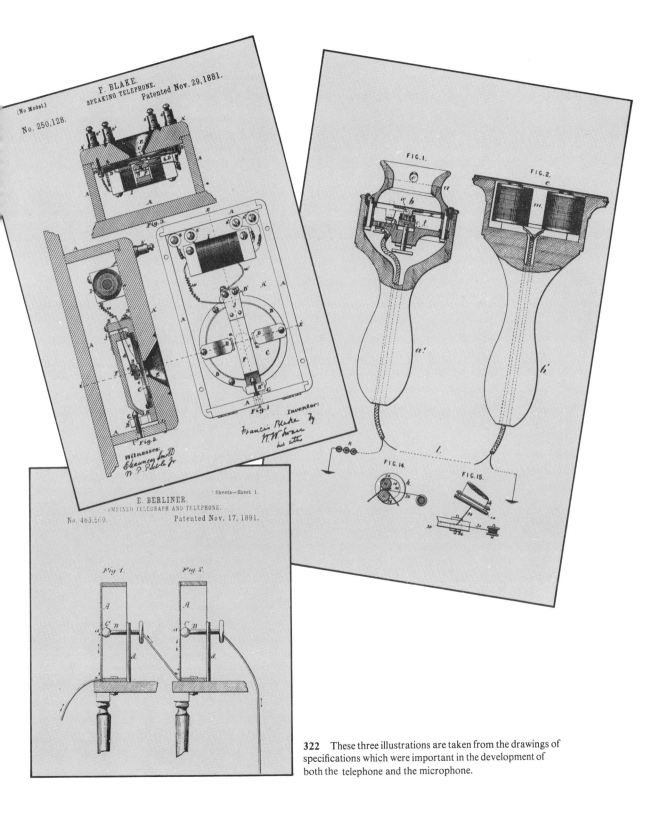

322 These three illustrations are taken from the drawings of specifications which were important in the development of both the telephone and the microphone.

222,604.   Television   apparatus.
BAIRD, J. L., The Lodge, Helensburgh, Dumbartonshire, and DAY, W. E. L., 15, Cholmeley Park, Highgate, London. July 26, 1923, No. 19173. [Class 40 (iii).]

Areas of the picture or scene to be transmitted are projected successively on a light-sensitive cell E, and the variations in the current in the cell circuit due to the lights and shades of the picture &c. are utilized to light at the receiving station a succession of lamps K arranged to form a screen, the differing intensities of illumination of the lamps constituting the picture to be viewed. An image C of the picture &c. A is projected by a lens B on a rotary disc D provided with a spiral series of equidistant holes M. Each hole in turn thus traverses a strip of the image and allows light corresponding to successive areas of the strip to fall on the cell E. At the receiving station, an arm G, moving synchronously with the disc D, moves over a series of contacts H to complete the circuits of the successive lamps.

324   From January 1889 to February 1931 the abridgments were published in numerical order in the *Illustrated Official Journal (Patents)*. Thus for this period there are available this set of abridgments in numerical order and the subject arranged set. On the left is a copy of the whole of the abridgment for the first of the two Baird specifications noted. The figures shown below were those from Zworykin's specification 290,245 which were copied with the abridgment for that patent.

July 4, 1928]   **THE ILLUSTRATED OFFICIAL JOURNAL (PATENTS).**

the same source and the speeds of the cylinders 14 are adjusted manually by means of brakes

18, 19.   In the modification in which use is made only of the parts 57, 58, Fig. 7, of the

spiral traced out by the exploring pencil, the time intervals corresponding to the remaining   parts of the spiral may be occupied by synchronising signals.

323 Telephone exchange system
940,016 (1963)   Improvements in or relating to automatic telecommunication systems.
POSTMASTER GENERAL and a number of private companies

This describes the first electronic exchange system to be installed by the GPO at the Highgate Wood exchange. It was not an immediate success and further development was required.

324 Television
222,604 (1924)   A system of transmitting views, portraits, and scenes by telegraphy or wireless telegraphy. BAIRD, JAMES LOGIE and DAY, WILFRID
290,245 (1928)   Improvements in optical systems for the scanning or recombining devices of picture transmission or television systems.
ZWORYKIN, VLADIMIR KOSMA, Westinghouse Electric and Manufacturing Company
United States 1,773,980 (1930)   Television system. FARNSWORTH, PHILO T., Television Laboratories Inc.

Baird was the first to show the possibility of developing a television system. His mechanical system was not, however, capable of refinement to a sufficiently sensitive system for widespread acceptance. Zworykin and Farnsworth conceived the concept of an electronically controlled system. Farnsworth had formulated the basic outline of his system at a very young age and was fortunate in being able to persuade a financial consortium to finance his research. Working in a small laboratory with a limited amount of assistance he overcame difficulties as they arose. His first patent, which is the one noted above, included a description of his image dissector tube which constituted his most important inventive contribution. Zworykin's major contribution was the Iconoscope, a device that transmits television images quickly and effectively. The first broadcasts were in England, first with the crude Baird equipment and then with an electronic system. ELECTRICAL AND MUSICAL INDUSTRIES LTD developed the Zworykin system in Great Britain and, with their own research group, which included A. D. BLUMLEIN and P. W. WILLIAMS,

produced a version of the Iconoscope, known as the Emitron, which was used in the regular BBC broadcasts initiated in 1936. There was very soon an exchange of patent licences between E.M.I. in England and RADIO CORPORATION OF AMERICA who held the rights to the Zworykin patents. See also under COLOUR TELEVISION.

325 Television recording
798,927–30 (1958)   Improvements in and relating to broad band magnetic tape recording and reproduction. GINSBURG, CHARLES PAUSON; HENDERSON, SHELBY FRANCIS; DOLBY, RAY MILTON and ANDERSON, CHARLES EDWARD, Ampex Corporation

These patents disclose the system marketed under the trade mark *Videotape* which is the basis of most television recording methods in use today.

326 Tenderising meat
913,202 (1962)   Improvements in or relating to meat products. BEUK, JACK FRANK; SAVICH, ALFRED LEO; GOESER, PAUL ALBERT and HOGAN, JOHN MATTHEW, Swift and Company

This patent is for a process for making all parts of meat carcasses uniformly tender when cooked. Activated enzymes are injected before slaughter. Slaughtering then takes place after a controlled period long enough to allow the enzyme to be distributed evenly throughout the animal but not long enough for the enzyme to be eliminated by the animal's body processes. At first the application for this patent was refused on the grounds that since it proposes the use of the normal living processes of the animal's body to distribute the enzyme material it was not a 'manner of new manufacture' as required by the patent Acts. This decision was reversed by the High Court and the patent allowed to proceed to grant.

327 Tennis and tennis balls
685/1874   A new and improved portable court for playing the ancient game of tennis. WINGFIELD, WALTER CLOPTON
4,623/1914   Improvements in methods and moulding devices for making rubber articles.

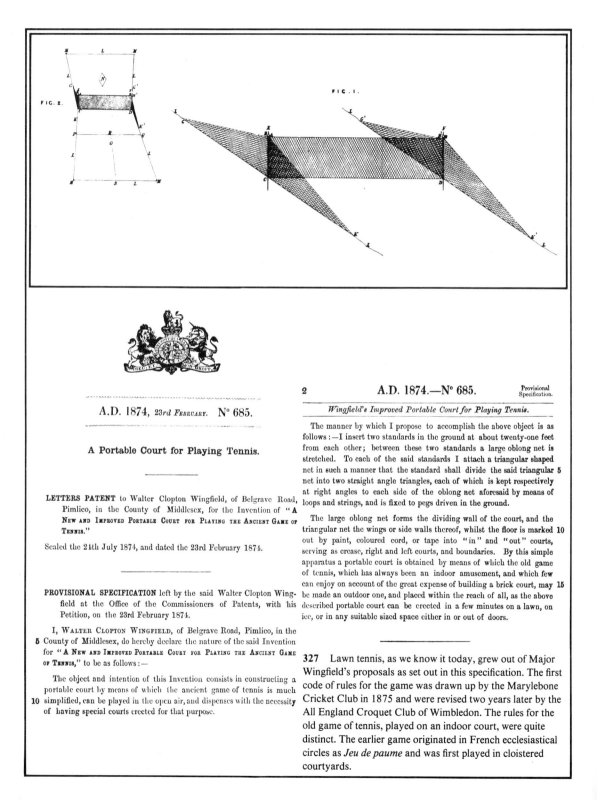

FIG. 2.

FIG. 1.

A.D. 1874, 23rd FEBRUARY.   N° 685.

A Portable Court for Playing Tennis.

LETTERS PATENT to Walter Clopton Wingfield, of Belgrave Road, Pimlico, in the County of Middlesex, for the Invention of "A NEW AND IMPROVED PORTABLE COURT FOR PLAYING THE ANCIENT GAME OF TENNIS."

Sealed the 24th July 1874, and dated the 23rd February 1874.

PROVISIONAL SPECIFICATION left by the said Walter Clopton Wingfield at the Office of the Commissioners of Patents, with his Petition, on the 23rd February 1874.

I, WALTER CLOPTON WINGFIELD, of Belgrave Road, Pimlico, in the
5 County of Middlesex, do hereby declare the nature of the said Invention for "A NEW AND IMPROVED PORTABLE COURT FOR PLAYING THE ANCIENT GAME OF TENNIS," to be as follows:—

The object and intention of this Invention consists in constructing a portable court by means of which the ancient game of tennis is much
10 simplified, can be played in the open air, and dispenses with the necessity of having special courts erected for that purpose.

2   A.D. 1874.—N° 685.   Provisional
                                                                Specification.

Wingfield's Improved Portable Court for Playing Tennis.

The manner by which I propose to accomplish the above object is as follows:—I insert two standards in the ground at about twenty-one feet from each other; between these two standards a large oblong net is stretched. To each of the said standards I attach a triangular shaped net in such a manner that the standard shall divide the said triangular 5 net into two straight angle triangles, each of which is kept respectively at right angles to each side of the oblong net aforesaid by means of loops and strings, and is fixed to pegs driven in the ground.

The large oblong net forms the dividing wall of the court, and the triangular net the wings or side walls thereof, whilst the floor is marked 10 out by paint, coloured cord, or tape into "in" and "out" courts, serving as crease, right and left courts, and boundaries. By this simple apparatus a portable court is obtained by means of which the old game of tennis, which has always been an indoor amusement, and which few can enjoy on account of the great expense of building a brick court, may 15 be made an outdoor one, and placed within the reach of all, as the above described portable court can be erected in a few minutes on a lawn, on ice, or in any suitable sized space either in or out of doors.

327 Lawn tennis, as we know it today, grew out of Major Wingfield's proposals as set out in this specification. The first code of rules for the game was drawn up by the Marylebone Cricket Club in 1875 and were revised two years later by the All England Croquet Club of Wimbledon. The rules for the old game of tennis, played on an indoor court, were quite distinct. The earlier game originated in French ecclesiastical circles as *Jeu de paume* and was first played in cloistered courtyards.

ROSENFELD, RALPH HENRY and ROBERTS, FRED THOMAS

Wingfield's patent relates to a method of erecting a net so that a game that had previously been played only indoors in specially built courts could be played in the open and is often quoted as being the origin of the presentday game of lawn tennis. The early tennis balls were made from sheet rubber formed into a roughly spherical shape before vulcanisation. W. H. COX applied for a patent for the process in 1897, but this was never published. A later machine giving similar results is described in 229,004 (1925), GRAY, CHRISTIAN HAMILTON. The Rosenfeld patent above describes a method in which two semicured hemispheres of rubber are joined. This method gave much more accurate results.

328 Terylene
578,079 and 579,462 (1946) Improvements relating to the manufacture of high polymeric substances. WHINFIELD, JOHN REX and DICKSON, JAMES TENNANT, Calico Printers Association

These patents were applied for in 1941 but the war delayed their acceptance. ICI undertook the commercial development in England and Du Pont in the United States. Terylene was the first all-British truly synthetic fibre.

329 Testing electrical components
886,660 (1962) Improvements in arrangements for automatically performing tests on electrical components. CLARK, THOMAS HAROLD and others. British Telecommunications Research Limited

An apparatus for taking electrical components from magazines and testing them, before presenting them to be assembled on printed circuit boards. For an apparatus for electric circuit testing, see 886,910 (1962) MINNEAPOLIS-HONEYWELL REGULATOR COMPANY

330 Thermos flask
4,421/1904 Improvements in or relating to bottles and the like. BURGER, REINHOLD

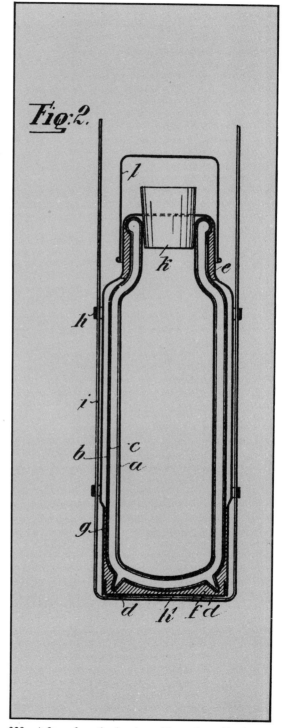

330 A figure from the drawings of Burger's specification 4,421.

331   Thermostat
6,014 (1830)   An apparatus for regulating temperature in vaporisation, distillation and other processes. URE, ANDREW

The above patent is listed as significant for no better reason than that Ure is credited with the coining of the term 'thermostat'. The specification describes several types of heat governor including one incorporating a bimetallic element. This was by no means the first bimetallic device, JOHN HARRISON used a bimetallic 'gridiron' clock pendulum to adjust for effects of changes in temperature in his chronometer of 1726. It is not even sure that Ure's bimetallic strips would have been practicable. The first thermostat to come into widespread use was one patented by CHARLES EDWARD HEARSON, 5,141/1881, designed for use in poultry incubators. This used a capsule filled with liquid which boiled and caused the capsule to expand and activate a control lever at the required temperature. Developments in thermostats have been many and varied since that time. Among these are: those using corrugated metal bellows, 11,548/1903, FULTON, W. M. and later developments; those for the control of electric furnaces introduced to this country by the CAMBRIDGE INSTRUMENT COMPANY, 13,524/1908 and 194,597 (1923), the first commercially succesful application to domestic ovens, the 'Regulo' introduced in the years 1922–23 by RADIATION LIMITED, and the first internal combustion engine thermostat developed by CAPTAIN MATT PAYNE, founder of the British Thermostat Co. Ltd, 435,568 (1935).

332   Threshing machine
1,645 (1788)   Machine, which may be worked by cattle, wind, water, or other power, for the purpose of separating corn from the straw. MEIKLE, ANDREW

This machine was the first to incorporate a threshing drum. Meikle was also granted a patent for a winnowing machine, 896 (1769), which was probably a copy or development of machines seen in Holland by his father JAMES MEIKLE.

333   Time-switch (clockwork)
2,435/1867   Improvements in apparatus for regulating the supply of gas to burners. THURGAR, WALTER CHRISTOPHER

The above patent was for switching gas light on and off. For street lamp switch, see 7,011/1897.

334   Time-switch for electricity
27,849/1904 and 27,290/1905   Improvements in timing devices for turning on and off gas, electric currents and other agents. HORSTMANN, ALBERT and others

335   Tin cans (for food preservation)
3,372/1810   Preserving animal and vegetable food. DURAND, PETER.

The original invention of the process must be credited to Frenchman NICOLAS APPERT who was awarded a prize in 1809.

336   Tracked vehicles
953 (1770)   Portable railway or artificial road to move along with any carriage to which it is applied. EDGWORTH, RICHARD
14,064/1904   Harvesting, etc., machine wheels. ROBERTS, DAVID

The first specification above was the first to suggest the concept while the second describes the first practical application to farm vehicles.

337   Tractors
253,566 (1926)   Apparatus for coupling agricultural implements to tractors and automatically regulating the depth of work. FERGUSON, HARRY

Describes the original concept of a hydraulic power lift linkage for coupling an implement to a tractor. All modern tractors have such a linkage which is often referred to as the 'Ferguson system'.

338   Transformer
5,201/1885   Improvements in induction apparatus for transforming electric currents. ZIPERNOWSKI, CARL; DERI, MAX and BLATHY, TYTUS OTTO

NOW KNOW YE, that in compliance with the said proviso, and in pursuance of the said Statute, I, the said Richard Lovell Edgeworth, do hereby declare that my said new Invention is composed and made in manner following, that is to say :—

The great advantages of wooden railways or artifical roads for the wheels 10 of heavy carriages to roll upon are so obvious, in removing such obstructions as the circumference of wheels would otherwise meet with in soft or rugged roads, that it would be necessary to enumerate them. It has been the great expence of laying such railways for carriages in general that has hitherto deprived the publick of the great advantage that might otherwise be obtained 1: by this principle. My Invention consists in making portable railways to wheel carriages, so that several pieces of wood are connected to the carriage, which it moves in regular succession in such manner that a sufficient length of railing is constantly at rest for the wheels to roll upon, and that when the wheels have nearly approached the extremity of this part of the railway their 2(motion shall lay down a fresh length of rail in front, the weight of which in its descent shall assist in raising such part of the rail as the wheels have already passed over, and thus the pieces of wood which are taken up in the rear are in succession laid in the front, so as to furnish constantly a railway for the wheels to roll upon.   2:

In witness whereof, I have hereunto set my hand and seal, this Twentieth day of April, in the year of our Lord One thousand seven hundred and seventy.

RICHARD LOVELL EDGEWORTH. (L.S.)

Sealed and delivered (being first duly   3
    stampt), in the presence of

THO⁵ HANDLEY.
THO⁵ YOUNGMAN.

336 This specification describes an idea that was to lead to a whole series of tracked conveyances from agricultural tractors, road construction vehicles and armoured tanks.

There is difference of opinion concerning to whom the credit for the invention of the early transformers should be awarded. This patent, by three Hungarians, was the first to propose an induction apparatus using a magnet of closed-in shape, i.e. circular or polygonal. Earlier versions, the best known being those of the Frenchmen Gaulard and Gibbs (see under ELECTRICITY DISTRIBUTION), used bar magnets.

339   Transistors
694,021 (1953)   Apparatus employing bodies of semiconducting material. WESTERN ELECTRIC COMPANY INC.
700,231 (1953)   Improvements in electrical semiconducting devices and systems utilising them. WESTERN ELECTRIC COMPANY INC.

Three men, JOHN BARDEEN, WALTER BRATTAIN and WILLIAM SHOCKLEY, were credited with the discovery of the first workable transistors by the award of the Nobel Prize in physics for 1956. The first announcement of the breakthrough to the production of a practical transistor was made in a short letter to *The Physical Review* in July 1948.

340   Transmission
3,388/1877   Improved means for transmitting motive power to the driving wheels of velocipedes. STARLEY, JAMES

341   Tube making
7,707 (1838)   Manufacture of brass and copper tubing. GREEN, CHARLES
This concerns the manufacture of brass and copper tubes by casting the lead in a mould with a steel bar as the core. See also 8,838/1841, 13,752/1851.

342   Tubular plastic film
741,963 (1955)   Improvements in or relating to the production of films. BUTEAUX, RICHARD HAROLD BARCLAY and CORNFORTH, WITHERINGTON, Imperial Chemical Industries

A method of producing tubular plastic films, i.e. of polyethylene, which have had many applications in the packing industry, for instance for the packing of frozen chickens.

343   Tumbler lock
1,430 (1784)   Lock for doors, cabinets, and other things on which locks are used (without wheels or ward). BRAMAH, JOSEPH

A successful tumbler lock had been patented a few years earlier by ROBERT BARRON, 1,200 (1778). Bramah's lock was a significant improvement but required an accuracy of workmanship in its manufacture that was rare at the time. It was about this time that Bramah recruited on to his staff the young HENRY MAUDSLAY. Together they set about devising suitable machine tools to manufacture the locks. These tools contained many features which were significant steps forward in machine tool development. They were not patented but were kept in secret workshops. Bramah and Maudslay together are often credited with being the originators of the application of the slide rest to bar lathes.

344   Tuning fork watch
761,609 (1956)   Electronic device for the operation of a time-piece movement. BULOVA WATCH COMPANY, INC.

This is the earliest watch controlled by a tuning fork and maintained by a transistor circuit. It was by no means the first use of a tuning fork for timekeeping. WILLIAM HENRY ECCLES in 1919 had used a thermionic valve circuit to maintain a tuning fork in continuous vibration and in 1924 the Post Office used an elinvar fork for accurate time keeping. Even before this tuning forks had been used as governors for electric motors and as early as 1869 D'ARLINCOURT proposed the use of tuning forks, 1,920/1869, as a speed control for facsimile equipment. The Bulova watch was marketed as the 'Acutron' and guaranteed to keep time with an accuracy of one minute per month when worn on the wrist.

345   Turbine engines
6,735/1884   Improvements in rotary motors actuated by elastic fluid pressure and applicable also as pumps. PARSONS, CHARLES

This steam turbine was quickly applied to the generation of electricity and ship propulsion. The patent was given an extension of five years after a

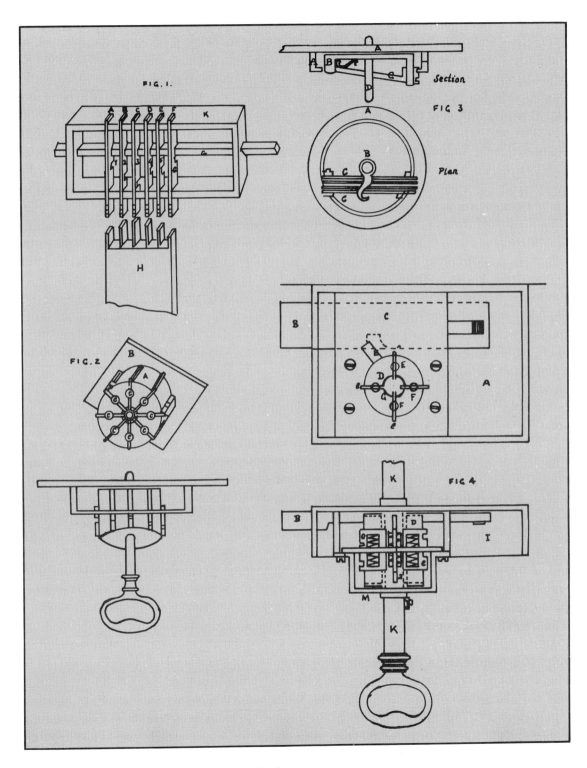

343 The drawings from Bramah's tumbler lock specification.

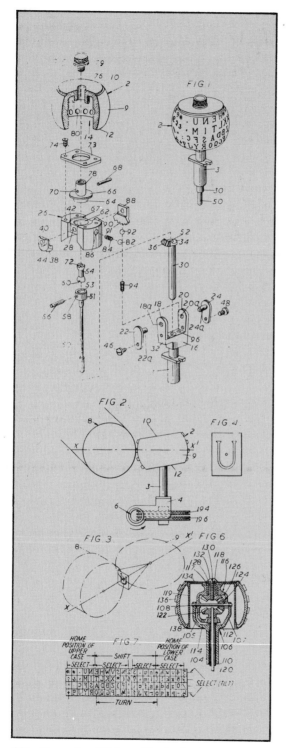

346   Figures from IBMs specification for the golf ball head.

report by Lord Kelvin had prophesied that the turbine would supersede the steam engine.

346   Typewriter

2,418/1879   Improvements in typewriting machines. Lake, W. R. for SHOLES, FREDERICK and PAGE, WILLIAM CURREY. United States patent was 207,559 (1878) to SHOLES, CHRISTOPHER LATHAM
5,789/1883   Typewriting machines. Lake, William Robert for HAMMOND, JAMES BARTLETT. The United States patent was 224,183 (1878).

The first typewriter patent was 395 (1714), MILL, HENRY. He never succeeded in perfecting his invention and his work died with him. The next developments of any significance were in France and Italy where work was aimed at producing a machine to aid the blind. One such machine is described in French patent 3,748 (1833), PROGIN, XAVIER. It was the Scholes' Patent above backed by the business acumen of JAMES DENSMORE of E. Remington and Sons that led to the introduction of typewriters on a commercial scale into ordinary offices together with the steady infiltration of women into business life. The Hammond patent above describes a typewriter using a swinging sector mechanism. This has been perpetuated in what is now known as the 'Varityper' and has the advantage that it is easy for the operator to change sectors and produce work in a variety of typefaces. A more modern development permitting the same facility is the 'golf-ball' head typewriter introduced with the IBM 72 in 1961, see 842,328 (1960).

347   Tyres, pneumatic

10,607/1888   An improvement in the tyres or wheels for bicycles, tricycles, or other road tyres. DUNLOP, JOHN BOYD
10,990 (1845)   Carriage wheels. THOMSON, ROBERT WILLIAM

Thomson's patent was bought up by a firm of coachbuilders who fitted such tyres to broughams but were unable to capitalise on the invention due to a general lack of interest. The idea was then forgotten until Dunlop reinvented it when seeking to improve his son's bicycle. As a tyre for cycling it was much more successful and soon showed its value when used for racing bicycles. The early

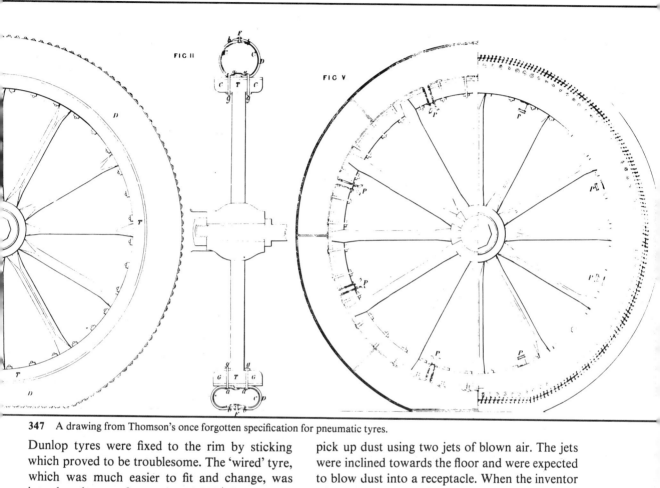

347 A drawing from Thomson's once forgotten specification for pneumatic tyres.

Dunlop tyres were fixed to the rim by sticking which proved to be troublesome. The 'wired' tyre, which was much easier to fit and change, was introduced soon afterwards, 14,563/1890 WELCH, C. K.

348 Universal joint
8,981 (1841) Machinery for propelling vessels on water—partly applicable to steam engines on land. BODMER, JOHN GEORGE

349 Vacuum cleaner
17,433/1901 Improvements relating to the extraction of dust from carpets and other material. BOOTH, HUBERT CECIL. See fig. 4, p. 15.

Booth first became interested in the removal of dust from carpets in 1901 when he was invited to the Empire Music Hall to see an American inventor demonstrate a machine that purported to pick up dust using two jets of blown air. The jets were inclined towards the floor and were expected to blow dust into a receptacle. When the inventor who was demonstrating reacted unfavourably to Booth's suggestion that sucking at the dust would be more effective Booth went away determined to investigate possibilities himself. He reports how he almost choked himself in a preliminary experiment by sucking with his mouth against the plush seat of a London restaurant. He sought financial support for further work and then set about specifying the mechanical necessities for a successful cleaner. The first machine he constructed was immediately successful. One of its early tasks was to clean the great blue coronation carpet under the throne of Westminster Abbey for Edward VII's coronation. There were about thirty patents for suction cleaning devices before Booth's but none of them had been developed sufficiently to produce a workable cleaner.

350   Vaseline

1,012/1874   Treating hydrocarbon oils, etc. CHESEBROUGH, WILLIAM HENRY. See also United States 127,568 (1872), CHESEBROUGH, ROBERT A. The trademark 'Vaseline' is a mixture of the German word *Wasser* and the Greek *elaion* meaning oil. Hence 'water-oil'. Crude petroleum oil had been used for medicinal purposes for hundreds of years but this process was the first that enabled it to be produced so that the full value of the oil was retained without any harmful impurities.

351   Velcro fastener

721,338 (1955)   Improvements in or relating to a method and a device for producing a velvet-type fabric. VELCRO, S. A.

One of several patents relating to fabrics with a surface of hooked pile threads which interlock when two such surfaces are pressed together.

352   Venetian blinds

945 (1769)   Window blinds. BEVAN, EDWARD

353   Vulcanisation

United States 3,633 (1844)   India-rubber fabrics. GOODYEAR, CHARLES
9,952 (1843)   Preparing caoutchouc in combination with other substances. HANCOCK, THOMAS
777/1858   The manufacture of improved belting for machinery and other purposes. PARMELEE, SPENCER THOMAS

There is little doubt that Goodyear must be given the credit for originating the process of vulcanisation that gave birth to the rapid growth of the rubber industry. Both the Goodyear and Hancock patents describe methods, involving mixtures of rubber and sulphur, that are in many ways alike. Goodyear perfected his method before Hancock who recalls having been shown samples of rubber vulcanised by the Goodyear process before he began the final experiments that led to his own success. There had been earlier attempts at vulcanisation using rubber-sulphur mixtures but none with quite the same dramatic success. The most notable of the earlier attempts was that of JAN VAN GEUNS of Holland who had been producing what may well have been vulcanised rubber from about 1836 and manufacturing sheets, air cushions and rubber-lined fire hose. In 1838 NATHANIEL HAYWARD had exposed mixtures of sulphur and rubber to solar radiation to give better 'drying' qualities, United States 1,090 (1839). It was Hayward who first interested Goodyear in mixtures of rubber and sulphur. Parmelee's patent describes a process which led to the development of the 'rotacure' method of continuous vulcanisation and which was applied to the making of rubber belting for conveyors.

354   Wankel engine

791,689 (1958)   An improved high-compression rotary piston machine with internal axes. WANKEL, FELIX, NSU WERKE A.G.

The first specification relating to a rotary-piston machine comprising the essential features of what is now known as the Wankel engine. The specification describes the basis geometry of the piston and casing profiles, which is of major importance and common to all Wankel-type engines. Other patents of note are 886,307 (1962), 892,476 (1962) and 893,938 (1962).

355   Water closet

1,105/1755   Water closet. CUMMING, ALEXANDER

The water closet had been invented and worked out in detail in 1596 by SIR JOHN HARINGTON. This patent embodies all the required elements—the supply, cistern, the soil pan interconnected with the flush and a pull-up handle, but it had the defect that the exit to the soil pan was closed by a sluice valve. Significant improvements in the valve arrangements were made by JOSEPH BRAMAH, 1,177 (1778) and 1,402 (1783). The manufacture and installation of water closets formed a substantial part of Bramah's business despite his many other inventions.

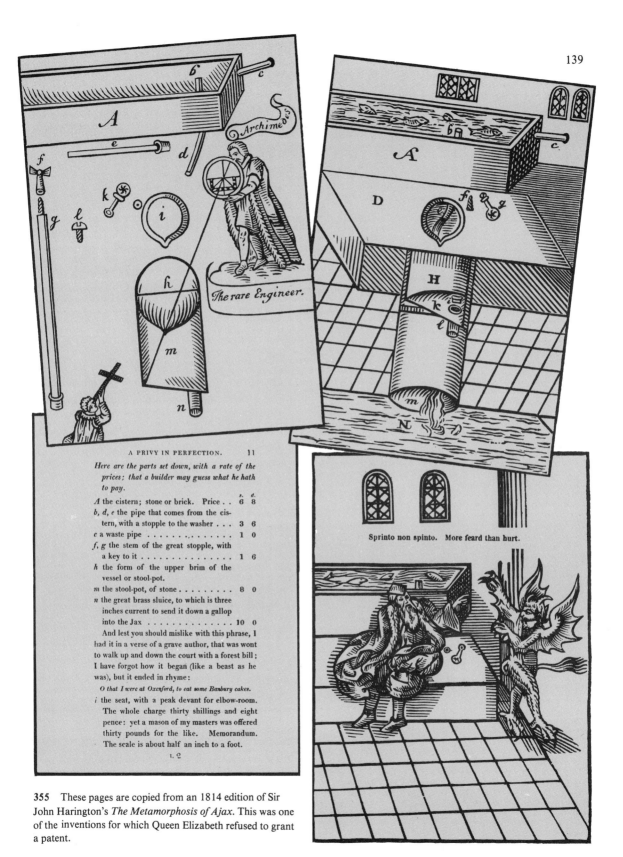

A PRIVY IN PERFECTION. 11

Here are the parts set down, with a rate of the prices; that a builder may guess what he hath to pay.

	s.	d.
A the cistern; stone or brick. Price	6	8
b, d, e the pipe that comes from the cistern, with a stopple to the washer	3	6
c a waste pipe	1	0
f, g the stem of the great stopple, with a key to it	1	6
h the form of the upper brim of the vessel or stool-pot.		
m the stool-pot, of stone	8	0
n the great brass sluice, to which is three inches current to send it down a gallop into the Jax	10	0

And lest you should mislike with this phrase, I had it in a verse of a grave author, that was wont to walk up and down the court with a forest bill; I have forgot how it began (like a beast as he was), but it ended in rhyme:

O that I were at Oxenford, to eat some Banbury cakes.

i the seat, with a peak devant for elbow-room. The whole charge thirty shillings and eight pence: yet a mason of my masters was offered thirty pounds for the like. Memorandum. The scale is about half an inch to a foot.

I. Q

Sprinto non spinto. More feard than hurt.

355 These pages are copied from an 1814 edition of Sir John Harington's *The Metamorphosis of Ajax*. This was one of the inventions for which Queen Elizabeth refused to grant a patent.

A godly father, sitting on a draught,
To do as need and nature hath us taught,

356   Watermarks

4,419 (1819)   Certain improvements in the manufacture of bank note paper for the prevention of forgery. CONGREVE, SIR WILLIAM

Plain watermarks had been in existence for some considerable time. This was the first description of a practical method of producing coloured watermarks. Although Congreve failed to get the Bank of England to adopt his plans in their entirety, this patent nevertheless marked a significant step forward.

357   Weaving loom

1,565 (1786)   New-invented weaving machine. CARTWRIGHT, EDMUND

This was Cartwright's second power loom patent and showed many improvements from the first, 1,470 (1785). A great improvement on this loom was the 'Wiperloom' patented by ROBERT MILLER, 2,122 (1796). An important step in loom development came with the introduction of automatic weft supply, 6,579 (1834), JOHNSON, THOMAS JOHN and REID, ARCHIBALD. See also SULZER LOOM, and NOZZLE LOOMS.

358   Windmills

615 (1745)   Self-regulating wind machine. LEE, EDMUND
1,628 (1787)   Regulator for furling and unfurling the sails of windmills when at work, for grinding corn and dressing flour and meal. MEAD, THOMAS
3,041 (1807)   Equalising the motion of windmill sails. CUBITT, WILLIAM

The origins of windmills stretch way back before the first patents. The earliest known reference to windmills in England dates back to 1191, while horizontal windmills were known in Arab countries, and possibly also in the Low Countries, as early as the tenth century. The above patents describe some of the improvements which were put into practice in the early years of patenting. The first relates to a fan-tail arrangement for keeping sails facing the wind. The fan-tail was set to face at right angles to the main sails of the windmill. When there was sufficient cross wind to turn the fan this operated a chain of gears to turn the main sails to

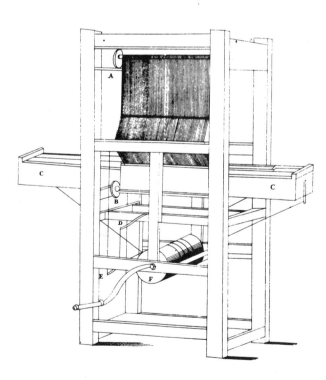

357   From Cartwright's first weaving specification.

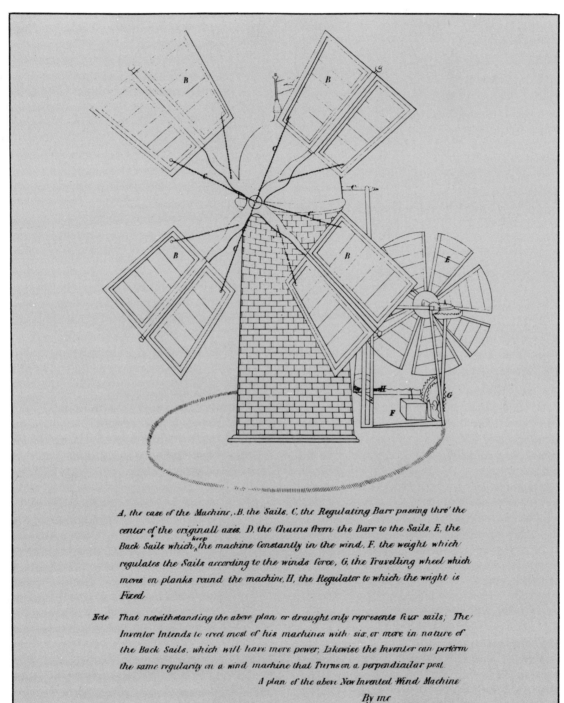

A, the case of the Machine,.B. the Sails, C. the Regulating Barr passing thre' the
center of the originall axes, D. the Chucns from the Barr to the Sails, E, the
Back Sails which keep the machine Constantly in the wind, F. the weight which
regulates the Sails according to the winds force, G, the Travelling wheel which
moves on planks round the machine, H. the Regulator to which the weight is
Fixed

Note   That notwithstanding the above plan or draught only represents four sails; The
Inventor Intends to erect most of his machines with six, or more in nature of
the Back Sails, which will have more power; Likewise the Inventor can perform
the same regularity on a wind machine that Turns on a perpendicular post.
          A plan of the above New Invented Wind Machine
                    By me
                        Edwd Lee.

358   An early example of feed-back control.

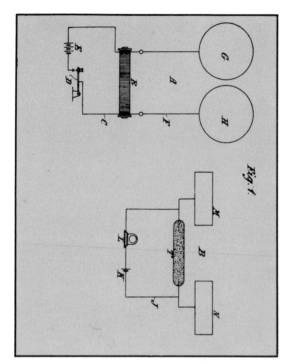

359 Figures from Marconi's first wireless specification.

the wind. The second patent proposed the use of a centrifugal force governor to drive a 'lift-tenter' mechanism. The lift-tenter itself had been invented in 1785 by ROBERT HILTON and served to apply force to prevent millstones moving apart from one another when driven at speed. The flyball governor was applied later by JAMES WATT and MATTHEW BOULTON to control the throttle valve in their continuously operating rotary steam engines. The last of the three patents concerns self-reefing sails. It is interesting to note that all three of these are examples of what is now fashionably called feedback control. This term was coined in the early part of the century by pioneers of radio and the concept has become basic to the new science of cybernetics which has been studied extensively since the 1930s and 1940s when biologists and economists began to note striking parallels between their own objects of study and the feedback control mechanisms of engineers. Much earlier examples of feedback control are known. An early water clock invented in the third century BC by a Greek named Ktesibios for the Egyptian King Ptolomy II is the earliest known example.

359 Wireless telegraphy

12,039/1896   Improvements in transmitting electrical impulses and signals, and in apparatus therefor. MARCONI, GUGLIELMO

Improvements are given in a later Marconi patent, 7,777/1900.

360 Xerography

672/767 (1952)   Electrophotography. CARLSON, CHESTER, Battelle Development Corporation
679,533 (1952)   Apparatus for producing photographs electrically. CARLSON, CHESTER, Battelle Development Corporation

The first of these two patents sets out the essential steps of the Xerox process. The second describes apparatus for performing the process including the photoconductive drum now used in copiers. The United States equivalent patents are 2,297,691 (1942) and 2,357,809 (1944).

361 Yale lock

United States Patent 161,727 (1875)
Combination locks. YALE, CHARLES O.

362 Yeast, growing on petroleum hydrocarbons

914,568 (1963)   Improvements in the production of yeasts. FILOSA, JEAN, British Petroleum Co. Ltd

This patent forms the basis of a large amount of subsequent study of methods of growing protein foods on petroleum hydrocarbons. The companion patent 914,567 (1963) is concerned with the recovery of the yeast from the hydrocarbon oils.

363 Zip fasteners

12,261 (1915)   Improvements in separable fasteners for articles of dress and for other purposes. SUNDBACK, GIDEON

Today's zip fasteners originate from the basic ideas described in this patent. It was the first form of the zip fastener to become commercially viable and was developed from an idea first patented by WHITCHURCH L. JUDSON, United States 504,037–8 (1893). The first British patent for a sliding-zip fastener was 24,782/1907, ARONSON, PETER ARON.

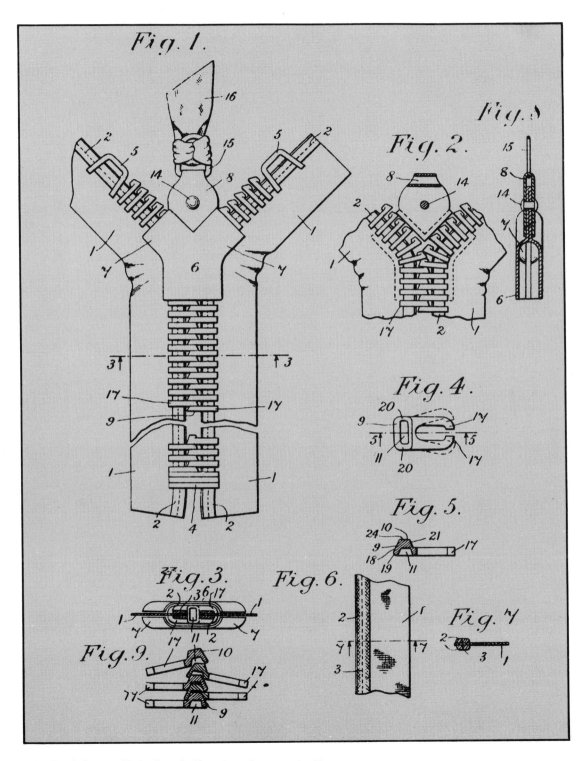

363 The zip fastener. No book on significant inventions can miss this out.

Name Index

Chronological Index

Subject Index